Heidelberg
Science
Library

Volume

21

Heidelberg
Science
Library

W. Guschlbauer

Nucleic Acid Structure

AN INTRODUCTION

Springer-Verlag
New York
Heidelberg
Berlin
1976

Wilhelm Guschlbauer
Service de Biochimie
Centre d'Études Nucléaires de Saclay
91190 Gif-sur-Yvette
France

Library of Congress Cataloging in Publication Data

Guschlbauer, Wilhelm.
    Nucleic acid structure.

    (Heidelberg science library; v. 21)
    Includes index.
    1. Nucleic acids. I. Title. II. Series.
QD433.G87   547'.596   75-11796

ISBN-13:978-0-387-90141-1   e-ISBN-13:978-1-4613-9397-9

DOI: 10.1007/978-1-4613-9397-9

To Marie-Pierre

To Florian, Theresia, and Willi

To "explain" an event means to trace or "reduce" its regularities to more general laws of nature. If the event occurs within a system consisting of several different parts, a knowledge of their form and function is, in any science, indispensable to successful reduction. While physicists are interested in structure only as a means to this end, biologists regard a knowledge of structure as an end in itself. The current belief that only quantitative procedures are scientific and that the description of structure is superfluous is a deplorable fallacy, dictated by the "technomorphic" thought-habits acquired by our culture when dealing preponderantly with inorganic matter.

Konrad Lorenz (1971)

Preface

Teaching a course on nucleic acid structure is a hazardous undertaking, especially if one has no continuous teaching obligations. I still have done it on several occasions in various French universities, when colleagues, suffering from administrative overwork and excessive teaching obligations, had asked me to do so. This was generally done with a pile of notes and a dozen slides, and I always regretted that no small, concise, specialized book on nucleic acid structure for students at the senior or beginning graduate level existed. Every year, the lecture notes became more and more voluminous, with some key reprints intermingled.

Everything changed when, in the spring of 1973, I received an invitation to teach such a course, under the UNESCO-OAS-Molecular Biology Program at the Universidad de Chile in Santiago during October 1973. I had accepted rather enthusiastically, but soon discovered that it would be necessary to produce a photocopied syllabus for the students. This was the first premanuscript of this book.

For nonscientific reasons, the course was first canceled and then postponed until December 1973. Nearly a year later, the course, in slightly amended form, was presented at the Lemonossow-State University in Moscow.

The manuscript was designed as a nonmathematical, voluntarily phenomenological introduction to nucleic acid structure, as a complement to the normally strongly enzymological approach in molecular biology. This book is not a monograph and has, therefore, no pretention to be a rigorous treatment of the physical chemistry of nucleic acids.

Besides the fact that our understanding of the structural constraints of nucleic acids is far from being precise, it is not very useful to charge students with such specialized problems and techniques as statistical thermodynamics, fast

kinetics, or quantum mechanical treatments, for which considerably more physical background is necessary than can normally be expected from a senior student in biochemistry. Also, many of the problems treated in this book are still evolving, and it is to be expected that new results will appear before the publication of the book. It is for this reason that protein–nucleic acid interactions, for instance, have been treated generally, rather than specifically.

On the other hand, certain new results have been included, despite the limited comprehension we have of them. This was done to demonstrate the "possibilities" of nucleic acids, their fine structure, and their future. For this reason, repeating sequences, restriction enzymes, and unusual polynucleotide complexes are treated.

It is certain that the choices of the subjects treated will be criticized. This is inevitable. The same is true for the references included. In general, and if available, review and symposium articles are cited preferentially. For those, who wish to deepen their understanding of the physical chemistry of nucleic acids, the recently published *Physical Chemistry of Nucleic Acids* by Bloomfield, Crothers, and Tinoco (Harper and Row, 1974) and the three-volume series *Basic Principles of Nucleic Acid Chemistry* edited by P. O. P. Ts'o (Academic Press, 1974) are recommended.

I shall not forget the many people who have been directly or indirectly involved in the elaboration of the manuscript. In the first place, there is Professor S. Litvak (now at the Université Bordeaux II) who invited me to Chile and suggested that I teach a structure-oriented course. Professor Z. Shabarova and Dr. E. S. Gromova (Moscow) further influenced the orientation of this book. Professor P. Cohen (Paris) generously made his lecture notes available. Professor P. Fromageot (Saclay) was a constant and generous supporter of the project. Professor A. E. V. Haschemeyer (New York) was a particularly thorough and severe commentator; and Professor S. Arnott (Purdue) read large parts of the manuscript and generously furnished X-ray fiber diffraction patterns and drawings of nucleic acid structures. Dr. G. Bernardi (Paris) was the first to suggest that I write a book and he and Dr. V. Vetterl (Brno) read the manuscript and made valuable comments and useful criticisms. My collaborators, Marie-Therese Sarocchi, Danielle

Helical columns have held a strong attraction for man for a long time, as shown by this medieval candelabra in the Church of Santa Maria in Cosmedin, in Rome.

Thiele, J. F. Chantot, P. Tougard, and Tran-Dinh S., have helped at different stages. Barbara L. Haas was the constant and thorough counsel in English style and spelling. To all of them, I express my deep gratitude. They are, obviously, not responsible for the shortcomings of the book.

Finally, my wife, Marie-Pierre, is to be thanked, for her patience and moral help all along these years.

Saclay, Spring 1975   Wilhelm Guschlbauer

Contents

1 Introduction

**1.1
History**

During the last 25 years, two classes of natural compounds, proteins and nucleic acids, have attracted most of the attention in what has been called molecular biology. This attention is due to the key role of these two classes of compounds in all cellular processes.

As we shall see, nucleic acids are the carriers of genetic information, whereas proteins are concerned with the execution of biological processes. The near invariability and exactness of genetic information is, in part, the result of the rather rigid structure of deoxyribonucleic acid (DNA) and the virtual chemical inertness of its structure under physiological conditions. Proteins, on the other hand, are quite adaptable to their different purposes because they can undergo a large variety of chemical interactions. Their structure is generally not rigid and can be changed by the action of outside agents (substrates, cofactors).

Two main purposes of nucleic acids can be noted:

1. Storage and faultless transmission of genetic information to progeny. This process, called replication, implies the exact copying of a DNA molecule to form two identical sets of cellular DNA.
2. Transmission and coding of this information into proteins. These processes, called transcription and translation, are mediated by a second class of nucleic acids, ribonucleic acid (RNA).

We shall see that the basic principle underlying all these processes is identical for all organisms and is based on a specific pairing of certain building blocks, the nucleic acid bases. We shall also see that most "errors," called mutations, are due to accidental physical or chemical changes in these building blocks, with resultant changes in their pairings.

The origin of our knowledge of hereditary processes and their chemistry came, independently and simultaneously, from the work of two pioneers: in 1863, Gregor Mendel discovered the basic laws of genetics, crossing garden peas of different phenotypes; in 1868, Friedrich Miescher found a macromolecular, phosphorous-containing substance in wound pus, which he called "Nuklein." It was 80 years before Avery, McLoed and McCarthy could show, in 1944, that the active principle behind these two discoveries was one and the same: deoxyribonucleic acid or DNA. It is worth noting that none of these discoveries was appreciated at their time.

It was only after World War II that the first significant advances were made, by the geneticists of the "phage group," centered around Delbrück and Luria, and by the "structuralists," of the Pauling and Astbury school. The x-ray work of British crystallographers and the chemistry of Chargaff's group led to the single most important discovery in biology of our century: the DNA double helix structure by Watson and Crick in 1953. This discovery provided, for the first time, a unified theory that took into account all the chemical, biological and physical data available at that time. The last 20 years have yielded a wealth of experimental data to confirm this theory, leading to the so-called "central dogma" of molecular biology.

**1.2
Cellular
Localization of
Nucleic Acids**

In eucaryotic cells, the largest cellular structure easily visualized in the light microscope is the nucleus. Within the nucleus the chromatin accounts for about 95% of cellular DNA. This roughly spherical structure possesses a triple membrane with numerous pores, which connect it with a cytoplasmic structure, the endoplasmic reticulum. Through these pores certain macromolecules, synthesized in the nucleus, e.g., messenger RNA (mRNA) can enter the cytoplasm where they become operational.

In the interior of the nucleoplasm, small spherical structures of high density can be localized. These are the nucleoli, in which most of the nuclear RNA occurs (about 20% of the total RNA).

On the endoplasmic reticulum, one finds the granular ribosomes, containing about equal quantities of RNA and protein, in aggregates that form polysomes or ergosomes.

Another cellular constituent, the mitochondrion, which is essentially responsible for the energetics of the cell, also contains RNA and DNA, but in small quantities (less than 1% of the protein by weight). It is probable that the mitochondrial DNA and RNA constitute the genetic equipment for the replication of these plasmids and control some of their protein synthesis. DNA and RNA are also found in chloroplasts and plants.

Bacteria and procaryotes, in general, have a less dense nuclear region; the procaryotic chromosomes are much more diffuse and are not delineated by a membrane at any stage of growth. All DNA (apparently in a single molecule) is concentrated in the chromosome. Polysomes seen in the bacterial cell are sometimes bound to the cell wall.

In viruses, the DNA (or RNA) is very densely packed into the capsides, which are always surrounded by protein.

2 Methods and Techniques

Surveys of many of the techniques in nucleic acid chemistry are found in some recent review articles (1–6). A new book by Haschemeyer and Haschemeyer (7) is an excellent text for nearly all the biophysical techniques. In this chapter, we deal briefly with four techniques that have found rather widespread use in recent years:

1. Optical activity [optical rotatory dispersion (ORD) and circular dichroism (CD)],
2. Nuclear magnetic resonance (NMR),
3. Ultracentrifugation
4. Fiber X-ray diffraction

In any spectroscopic method, i.e., whenever electromagnetic quanta interact with matter, the fundamental equation $E = h\nu$ is applicable, where h is Planck's constant, E the energy of the quantum of radiation, and ν its frequency. In any spectroscopic analysis, the absorption of electromagnetic radiation by a population of spectroscopically identical particles (molecules, atoms, electrons) as a function of frequency will appear as a bell-shaped, generally Gaussian, curve.

The three parameters that define such a spectral band are (a) intensity, which is the surface of the band; (b) position of the band peak or the frequency of the absorption maximum; and (c) breadth of the band, generally defined as the width at half-peak height.

2.1 Absorption and Optical Activity (ORD and CD)

Any substance the electrons of which absorb electromagnetic quanta in the range between 150 and 800 nm will have an ultraviolet (UV) or a visible absorption spectrum. Generally, the absorption bands will be Gaussian, as a function of frequency for a given transition. Coupling between transitions from neighboring chromophores may reduce the absorption observed at $\varepsilon_{\mathrm{max}}$ of the two isolated bands. This is the well-known hypochromism of nucleic acids, due to the interaction of the π-electrons of the bases (see Fig. 5-1).

For a substance to be optically active, it is necessary and sufficient to contain neither a plane nor a center of symmetry. In other words, the substance and its mirror image must not be superimposable.

Optical activity manifests itself in two ways: (a) by the rotation of the plane of polarization of linearly polarized light, due to the difference of refraction indices for the left- and right-handed components of linearly polarized light (see Fig. 2.1) and (b) by a difference in absorption between the left- and right-hand component of circularly polarized light at frequencies where absorption occurs, due to a difference in velocity between the two light components. In the first case, we have optical rotation, in the second, circular dichroism (CD) (Fig. 2.1). Both phenomena are produced by optically active molecules, the chromophores of which satisfy the requirement of asymmetry. This asymmetry need not be intrinsic, i.e., due to the chemical structure of the compound; it can be induced, e.g., through the binding of an optically inactive chromophore to a matrix to establish an asymmetric environment or through the polymerization of an optically inactive substance to form an asymmetric, optically active superstructure.

If an absorption band is optically active, a CD band with a shape similar to that of the absorption band will appear (Cotton effect). Optical rotatory dispersion (ORD) occurs

FIGURE 2.1. Simplified presentation of a dichroic band (Cotton effect). (a) Positive Cotton effect; (b) negative Cotton effect. Molar extinction coefficients for left- and right-hand polarized light components ① and ②; Refraction index of left- and right-hand polarized light components ③ and ④; Circular dichroism ⑤; Optical rotatory dispersion ⑥.

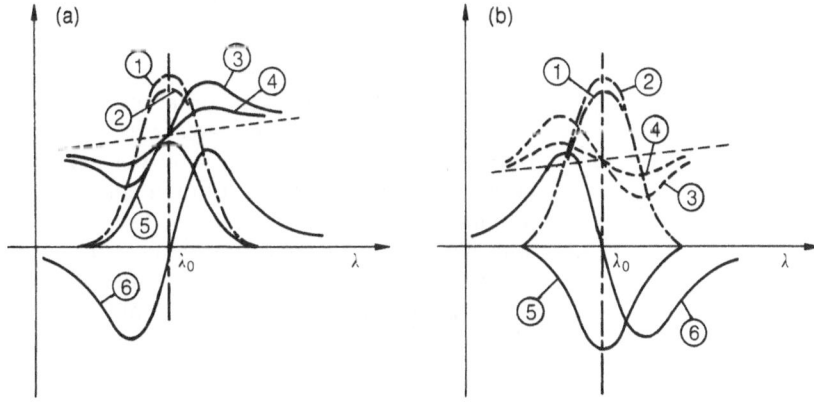

in regions outside the absorption band as a monotonous change of the rotatory power as a function of wavelength. In the region of the absorption band, an anomalous ORD will be found (Cotton effect) (Fig. 2.1). These spectra can be interconverted via the Kronig–Kramers transform; this amounts to the first derivative of the CD spectrum, which will be the ORD spectrum.

In principle, both CD and ORD yield similar information. There are, however, advantages and inconveniences to both techniques. The fact that ORD is observed over the whole spectral region permits the study of this phenomenon even in systems where the absorption band is outside the measurable range. Also, from ORD one can calculate using the Drude equation, certain useful optical parameters outside the experimentally accessible range. The advantage of CD is the simplicity of the spectra, which permits more exact band assignments (less band overlap). The disadvantage of ORD is the more complex band pattern, which can make it difficult to assign transitions precisely. But CD can apparently only be used within an optically active band being useless beyond this range (see Fig. 2.1).

The following units are used in ORD measurements:

Specific rotation: $\quad\quad\quad [\alpha] = \dfrac{100\alpha}{cd}$

Molar rotation: $\quad\quad\quad [m] = \dfrac{M[\alpha]}{100} = \dfrac{\alpha}{c_m l}$

Reduced molar rotation: $\quad [m'] = \dfrac{3[m]}{(n^2+2)}$

where α is the observed rotation in degrees, c the concentration in percent, d the cuvette path length in decimeters, l the cuvette path length in centimeters, M the molecular weight of the molecule (or monomer), and c_m the molar concentration.

Similarly, in CD the units used are:

Molar CD: $\quad\quad\quad\quad\quad\quad \Delta\varepsilon = \Delta A \cdot s/(c_m l)$
Molar ellipticity (θ or ψ): $\quad \theta = 3300\Delta\varepsilon$

where ΔA is the measured absorbance difference, s the sensitivity of the instrument (in absorbance units).

Since the nucleic acid bases are aromatic compounds (see Chapter 3), the analogy with benzene has led several authors to correlate the absorption spectra (which also determine CD) of these two series. The region between 200 and 300 nm shows four absorption bands in benzene and other aromatic compounds. The names B_{2u} (around 265 to 285 nm) and B_{1u} (around 230 to 260 nm) have been assigned. These bands are π-π^* bands and are polarized in the plane of the base. Below 220 nm, benzene possesses a doubly degenerate band, E_{1u}, which is also found in the purines and pyrimidines. These two bands (E_{1ua} and E_{1ub}) are also polarized in the plane of the base (π-π*) (see Fig. 3.3).

There exists the possibility of n-π* transitions, polarized perpendicular to the base plane. They can be distinguished by the classical McConnell criteria:

1. An electron-donating group (Cl, amino, methoxy) on an aromatic cycle displaces the π-π* transitions toward the red and the n-π* bands toward the blue.
2. Replacement of a nitrogen atom by a carbon has the opposite effect.
3. Hydrogen bond formation, protonation, and solvents with a high dielectric constant shift n-π* transitions toward the blue but do not affect the π-π* transitions. Deprotonation, breaking of hydrogen bonds, and aprotic solvents (hexane, etc.) induce red shifts.
4. In general, n-π* transitions have molar extinctions of a few hundred; π-π* band extinctions reach thousands and tens of thousands. This is not necessarily seen with CD, i.e. relatively large CD bands may be observed for n-π* transitions.

2.2 Nuclear Magnetic Resonance (NMR)

Nuclear magnetic resonance (NMR) spectroscopy is based on the absorption of radiofrequency electromagnetic radiation (of the order of 100 MHz) by atomic nuclei [mainly protons: proton magnetic resonance (PMR), but also ^{13}C, ^{15}N, ^{17}O, ^{19}F, ^{23}Na, ^{31}P, ^{25}Mg, etc.] in substances placed in a very strong magnetic field [10 to 100 kiloGauss (kG)]. In NMR, the band width is usually rather small (normally a few Hertz), compared with the absorbed frequency, and the absorption peaks are single rays for a given nucleus or a group of magnetically equivalent nuclei. In many cases the spectrum of nuclei in a single chemical group (e.g. methyl) shows a coupling of the spins with those of neighboring nuclei. From the disposition of the peaks, the spin-spin coupling constants can be evaluated. This is, however sometimes only possible with extensive calculations, requiring theoretical analysis and computers. Empirical relations have been deduced between the coupling constants of resonating nuclei and dihedral angles. The most commonly used form for couplings between vicinal protons is the Karplus Equation (8)

$$J_{ij} = J_0 . \cos^2 \phi_{ij} - 0.28 \text{ Hz}$$

where J_{ij} is the observed coupling constant in Hz, J_0 an empirical constant (which can vary between 8 and 16 Hz depending on the compound, the electronegativity of its substituents, and its geometry) and ϕ_{ij}, the dihedral angle in degrees. A modified form of this relation applicable to the ribose geometry has recently been proposed (9) in the form

$$J_{ij} = A \cos^2\phi_{ij} + B\cos\phi_{ij} + C$$

where $A = 10.5$ Hz, $B = -1.2$ Hz, and $C = 0$. Figure 2.2 shows the dependence of the coupling constant J_{ij} on the dihedral angle, using the two forms of the Karplus equation.

The principal parameters for NMR studies are chemical shifts, intensity (area) of absorption peaks, coupling constants, and relaxation times. Detailed interpretations of these spectroscopic parameters in NMR spectra depend on a number of suppositions, which, generally, have been verified.

1. The intensity of the absorption line is linearly proportional to the number of magnetically equivalent nuclei. For small molecules, comparison of intensities with chemical structure usually allows the unambiguous assignment of lines to individual atoms or groups of atoms in a molecule.

2. The resonance frequency (chemical shift, expressed as frequency relative to a standard) depends on the nature of the nucleus (^{1}H, ^{13}C, ^{19}F, etc.), the screening or shielding of the electrons surrounding the absorbing nucleus, and the spatial configuration of the nucleus. It may be influenced by permanent local fields, like unpaired electrons or ring currents from aromatic systems. Thus, protonation of a nitrogen of a nucleoside base will affect the neighboring protons greatly (Fig. 2.3). Titration of a phosphate group in a nucleotide will show displacements of the chemical shifts

FIGURE 2.2. The dependence of the proton-proton spin coupling constants J_{ij} on the dihedral angle ϕ_{ij}, using different forms of the Karplus equation.

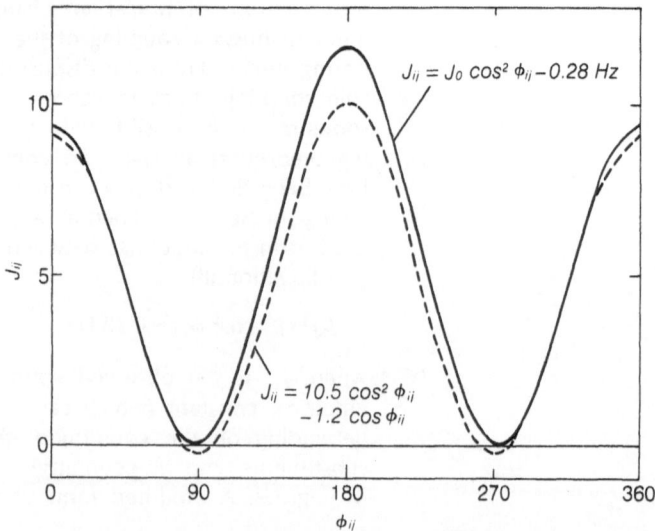

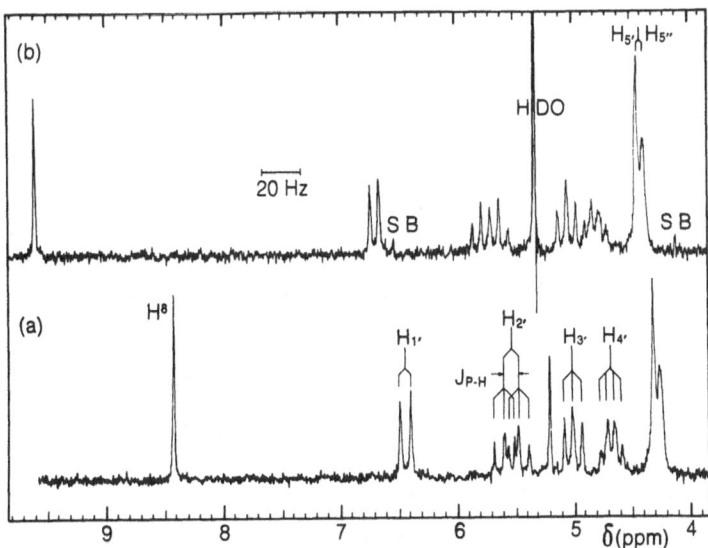

FIGURE 2.3. The 60-MHz spectra of Guo-2'-P at 25° C (0.14 M).
(a) pH=7.4; (b) pD=1.2. The large shift of the H_8 resonance is
due to protonation at N_7.

From Ref. 10. Reprinted with permission from *J. Amer. Chem. Soc.* © 1972
by the American Chemical Society.

of the adjacent sugar protons. Ring current effects have been
used to define the geometry of stacked oligonucleotides
and self-associating bases.

3. The band width will be determined by the rate of
atomic motion of the resonating nucleus, which, in turn, is
determined by the relaxation times t_1 and t_2. The larger
these relaxation times (i.e., the faster the motion), the nar-
rower the line widths.

4. Generally speaking, NMR permits the study of physical phenomena that involve single atoms and their imme-
diate neighborhood. Therefore, great changes in chemical
shifts can frequently be observed with minor changes in the
environment; on the other hand, otherwise major changes
in the system, which give rise to strong signals with other
techniques, may not change the NMR significantly. This is
particularly true if many absorbing nuclei are present, e.g.,
as in nucleic acids or proteins. It is for this reason that NMR
has, until now, been used mainly for small molecules, monomers, and oligomers. On the other hand, selective deu-
teration (deuterium does not absorb in the proton frequency
range) has been used effectively, but the technique requires
very extensive synthetic work. For the same reason deuterium oxide (D_2O) has to be used instead of H_2O as solvent.

For a more detailed discussion of the theory and the
biochemical applications of NMR, reviews (4, 11, 12) and
books (13–14) are available.

2.3 Ultracentrifugation

2.3.1 SEDIMENTATION IN HOMOGENEOUS MEDIA

At reasonably low concentrations (to avoid associations) and moderate rotor speeds, sedimentation of macromolecules follows a series of well-defined relations. For nucleic acids, UV optics may be conveniently used to measure the rate of sedimentation with time. At an angular velocity, ω, the distances x_1 and x_2 (in cm) traveled at times t_1 and t_2 (in seconds) are related to the observed sedimentation coefficient s_{obs}. This result may be normalized to correspond to the density of water at 20° as follows:

$$s_{obs} = \frac{1}{\omega^2 x}\frac{\partial x}{\partial t} = \frac{\ln{(x_2/x_1)}}{\omega^2(t_2 - t_1)}$$

$$s_{20,w} = \left[s_{obs}\frac{\eta_t}{\eta_{20}}\frac{\eta_s}{\eta_w}\frac{1 - \bar{v}\rho_{20,w}}{1 - \bar{v}\rho_{t,s}} \right]$$

where η_t and η_{20} are the viscosities of water at temperatures t and 20°C, η_s/η_w is the ratio of the solvent to water viscosity, $\bar{v}$ is the partial specific volume of the solute [$\bar{v}$ (DNA) = 0.55], and the ρ's are the densities of water and solvent at t and 20°C, respectively. The partial specific volume is $\bar{v} = 1/\rho_b$, where ρ_b is the buoyant density of the solute in the solvent. This yields

$$s_{20,w} = \left[s_{obs}\frac{\eta_t}{\eta_{20}}\left|\frac{\eta_s}{\eta_w}\right|\frac{\rho_b - \rho_{20,w}}{\rho_b - \rho_{t,s}} \right].$$

Since the viscosity of the solute is implicated in these equations, the sedimentation coefficient is not *a priori* a measurement of molecular weight; it is also influenced by shape, flexibility, solvatation, etc., of the solute. Therefore, standards of known molecular weight are generally run with the sample (or in a parallel experiment) to calibrate the measurements.

2.3.2 DENSITY GRADIENT CENTRIFUGATION (6)

In order to avoid many of the technical difficulties of sedimentation in homogeneous media (convection, associations, etc.), a continuous density gradient is often used. After a sedimentation run in a buffered gradient, established before the run in the centrifuge tube, a hole is punched in the bottom of the tube, and the solution collected drop by drop. Monitoring is by absorption, determination of radioactivity and enzyme activity, etc., and, frequently, a marker of known molecular weight is added to the solution.

Equilibrium density centrifugation in concentrated salt solutions (CsCl, Cs$_2$SO$_4$) (6,16), which will establish a density gradient during ultracentrifugation, exploits the differences in buoyant density of solutes. At equilibrium, the macromolecule under investigation will form a narrow band in the gradient at the density corresponding to its buoyant

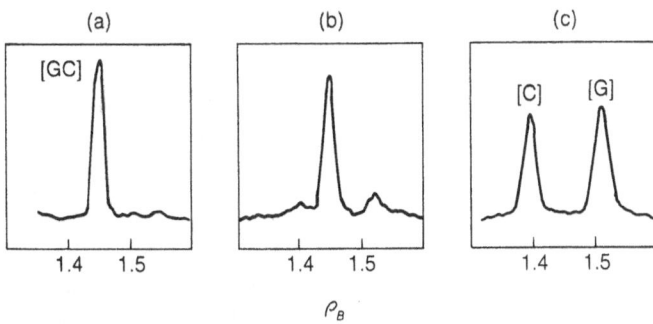

FIGURE 2.4. Density gradient centrifugation of poly(dG)·poly(dC) in Cs_2SO_4 at (a) pH<11.6; (b) pH=11.65; (c) pH>11.7.

density. Buoyant density is sensitive to variations in composition and/or structure, and may be used to estimate the base composition of DNA (17) from the relation

$$\rho_b = 1.66 + 0.098(G\text{-}C\%)$$

Complementary polynucleotide strands can be separated in alkaline Cs_2SO_4 or CsCl gradients (6,17) (Fig. 2.4).

2.4 X-Ray Fiber Diffraction

If a parallel monochromatic light beam is incident upon a slit, a Fraunhofer diffraction pattern will be observed through a lens in the focal plane of the lens. An infinite array of parallel slits (or a grating) will give a discrete spectrum of lines; the distance between these lines is inversely proportional to the distance between the slits, and the amplitude of the lines will depend on the slit width. This is an example of "reciprocal space."

If the slits are replaced by a regular array of atoms, e.g., a crystal or a periodic polymer, and if the light beam is replaced by a monochromatic X-ray beam, similar diffraction patterns will be obtained. Again, the distances in the pattern obtained will be inversely proportional to the distances in the atomic array. Since the wavelength of X-rays is in the Angström range, the diffracting units will be atoms.

A helix (Fig. 2.5) of pitch P, consisting of M scattering points, at distance p, in N turns of the helix of length C, will follow the equation

$$C = Mp = NP$$

Since C is the repeat distance, the spacings between the layer lines of the reciprocal space projected upon the diffraction pattern will be $1/C$. For a single discontinuous helix, the integral that defines the amplitude of diffraction will take finite values only at layer lines l/C distant from the origin, according to the selection rule

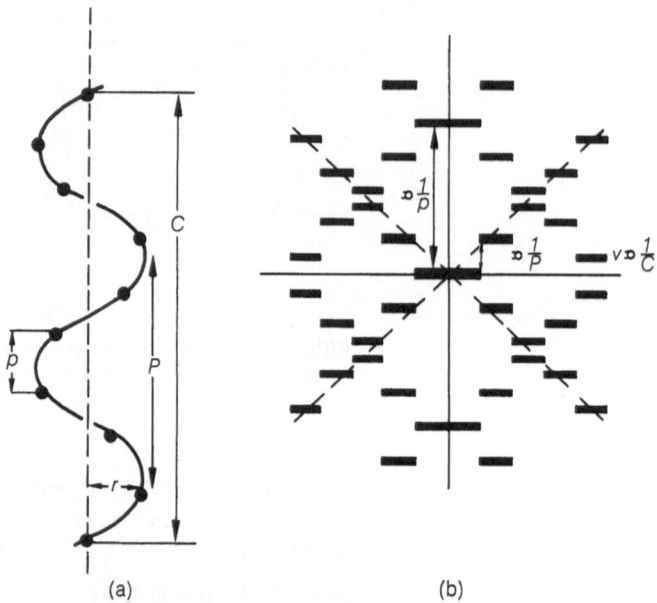

(a)   (b)

FIGURE 2.5. Discontinuous helix (a) with $M/N=9/2$ and its schematic X-ray diffraction pattern (b).

$$Z = l/C = m/P + n/p$$

where l, m, and n are integers.

Figure 2.6 shows a particularly simple fiber diffraction pattern. The meridian spot on the second layer line indicates that the repeat is every two scattering units and $M/N=2/1$.

For a short, but precise review, the book by Wilson (19) should be consulted.

References

1. Michelson, A. M., Massoulié, J. and Guschlbauer, W. (1967). *Prog. Nucleic Acid Res. Mol. Biol.*, **6**, 83–141.

2. Yang, J. T. and Samejima, T. (1969). *Prog. Nucleic Acid Res. Mol. Biol.*, **9**, 224–301.

3. Brahms, J. and Brahms, S. (1970). *Biol. Macromolecules*, **4**, 191–268.

4. Jardetzky, O. and Wade-Jardetzky, N. G. (1971). *Ann. Rev. Biochem.*, **40**, 605–34.

5. Arnott, S. (1970). *Prog. Biophys.*, **21**, 265–319.

6. Szybalski, W. (1968). *Methods Enzymol.*, **xii–2**, 330–60.

7. Haschemeyer, R. H. and Haschemeyer, A. E. V. (1972). *Proteins.* New York: Interscience.

8. Karplus, M. (1959). *J. Chem. Phys.*, **30**, 11–20.

9. Altona, C. and Sundaralingam, M. (1973). *J. Amer. Chem. Soc.*, **95**, 2333–44.

10. Tran-Dinh, S., Guéron, M., and Guschlbauer, W. (1972). *J. Amer. Chem. Soc.*, **94**, 7903–11.

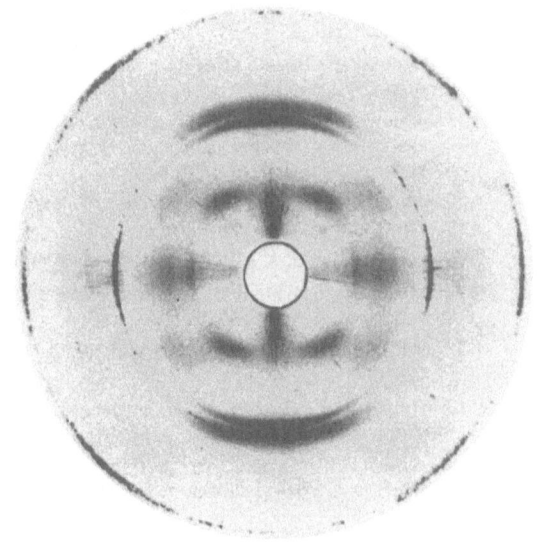

(a)

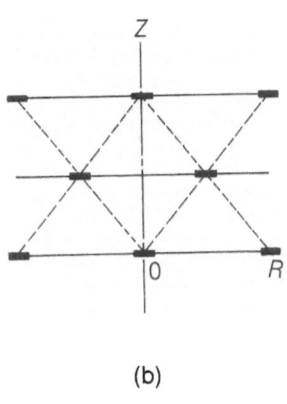

(b)

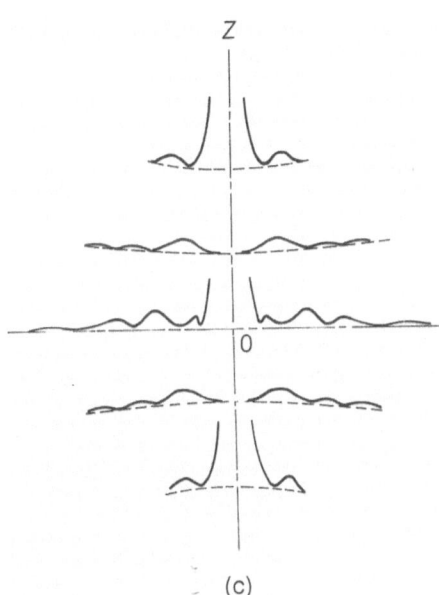

(c)

FIGURE 2.6. X-ray fiber diffraction pattern given by the Guo gel.
(a) Actual X-ray pattern; (b) schematic presentation; (c)
computed Fourier, using $M/N = 2/1$ and a fourfold symmetry.
(From Ref. 18.)

3 Chemistry and Enzymology of Nucleic Acids

Complete hydrolysis of nucleic acids (see Section 3.4) by chemical or enzymatic means liberates the nucleic acid "building blocks": phosphate, sugar (ribose or deoxyribose), and bases, in a 1:1:1 ratio. Depending on the conditions of the hydrolysis, larger compounds containing these building blocks are obtained (oligonucleotides).

3.1.1 THE BASES

All bases found in nucleic acids are derivatives of purine or pyridine. Only in some nucleoside antibiotics is an exchange of a carbon atom for a nitrogen atom, or vice versa, observed. The common bases are adenine: 6-amino-purine; guanine:2-amino-6-ketopurine; cytosine:2-keto-4-aminopyrimidine; uracil:2,4-diketopyrimidine; and thymine (in DNA): 2,4-diketo-5-methylpyrimidine (5-methyluracil). Minor components found in various DNA and RNA are 5-methylcytosine, 5-hydroxymethylcytosine, 5-hydroxymethyluracil, 6-methylaminopurine, 5,6-dihydrouracil (in tRNA) and numerous other methylated bases (in tRNA). The bases are quite resistant to oxidation but easily attacked by nucleophilic reagents. In particular, the positions meta to the nitrogen atoms have low electron densities. Substitutions by electron donors (amino, methyl, hydroxyl) facilitate nucleophilic substitution on the other atoms.

3.1.2 NUCLEOSIDES AND NUCLEOTIDES

If a base is linked through the N^1 of the pyrimidines or the N^9 of the purines with the $C_{1'}$ of the sugar moiety, (β-D-ribofuranose in RNA or β-D-2'-deoxyribofuranose in DNA), one obtains the nucleosides. This substitution will change the behavior of the base considerably: the solubility in water increases and that in aprotic solvents decreases. Considerable changes in *pK* intervene (Table 3.1). In the phosphate esters (nucleotides), the solubility in water increases further; the phosphate group itself is charged, with two ionization constants. Also, the absorption spectra of the bases are affected to a considerable degree by glycosylation. This introduces an asymmetric center ($C_{1'}$) into the molecule, which gives rise to optical activity, a property which can be usefully employed in identification and structural studies. In natural nucleic acids, the sugars (pentoses) are deoxyribose in DNA) and ribose (in RNA). Many synthetic nucleosides containing other sugar residues have also been prepared.

Table 3.1   Absorption and Ionization Characteristics
of Bases and Nucleosides

Compound	pH	λ (max)	ε (max)	pK	Site
Adenine	1	265.5	13.1		
(Ade)	7	260.5	13.4	4.1	N^1
	12	269.0	12.3	9.8	Amino
Adenosine	1	257.0	14.6		
(Ado)	7	260.0	14.9	3.4	N^1
	13	260.0	15.0	12.2	Amino
Guanine	1	275.5	7.35		
(Gua)	7	275.5	8.15	3.2	N^7
	13	273.0	8.0	9.6	N^1
Guanosine	1	256.5	12.2		
(Guo)	7	252.5	13.7	2.2	N^7
	12	258-266	11.3	9.2	
Cytosine	1	274.0	10.2		
(Cyt)	7	267.0	6.1	4.5	N^3
	12	282.0	7.9	12.2	Amino
Cytidine	1	280.0	13.4		
(Cyd)	7	271.0	9.1	4.4	N^3
	14	273.0	9.2	12.3	Amino
Uracil	1-7	259.0	8.2		
(Ura)	12	284.0	6.15	9.3	N^3
Uridine	2-7	262.0	10.1		
(Urd)	12	262.0	8.5	9.2	N^3
Thymine	2-7	264	7.9		
(Thy)	12	291.0	5.4	9.8	N^3
Thymidine	2-7	267.0	9.65		
(Thd)	12	267.0	7.4	9.8	N^3

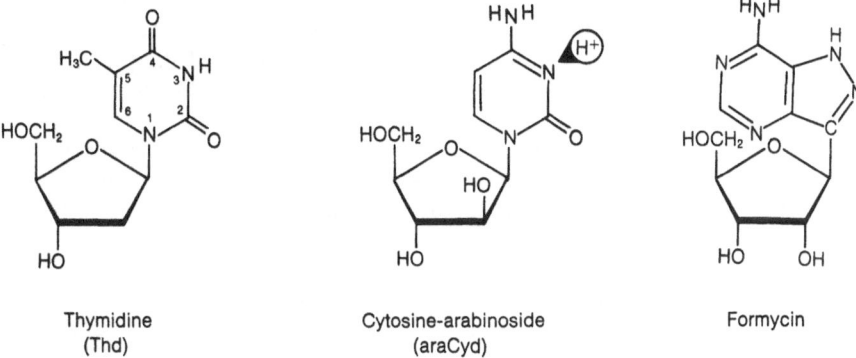
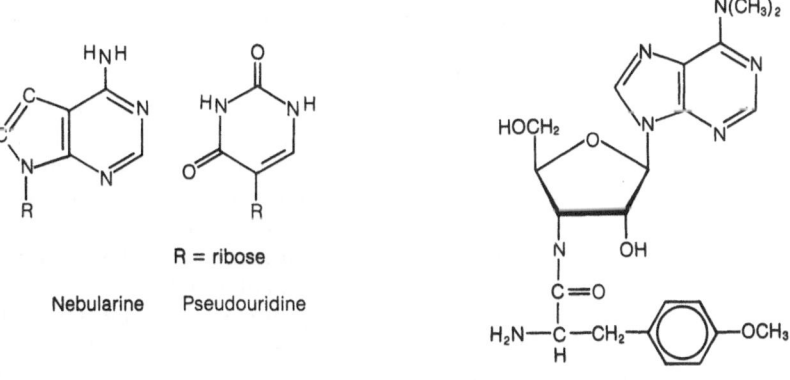

FIGURE 3.1. Some nucleosides and their derivatives.

Another naturally occurring sugar in nucleosides is arabinose. The arabinosides inhibit many nucleic acid enzymes and have widespread therapeutical applications. Some bases, nucleosides, and nucleotides, with indications of their protonation sites, are listed in Fig. 3.1. The abbreviations used are found in the Appendix.

3.1.3 STRUCTURE AND CONFORMATION OF NUCLEOSIDES AND NUCLEOTIDES

The structure and conformation of a nucleoside is determined by a variety of factors: ionization state, solvation, external influences (ions, cofactors, etc.), and temperature. All these parameters must be controlled so that conformation and, therefore, the behavior of these compounds, can be understood. We shall treat these points in some detail.

If one inspects the space-filling model of a nucleoside, say, guanosine (Guo), one notes a large degree of freedom of several groups. The base can rotate without much difficulty around the glycosidic bond (see Fig. 3.2). The definition of the torsion angle ϕ_{CN} (or χ_{CN}) in Fig. 3.2 (1) is given by the angle the plane of the base makes with the $C_{1'}$-$O_{1'}$ bond. The region in which the C^8 of the purines (or the C^6 of the pyrimidines) is above this bond is called *anti*, if C^8 or C^6 (see above) are above $H_{1'}$ the region is called *syn*.

Since optical activity measures perturbations in the neighborhood of the chromophore, this technique has been widely used to study nucleoside conformation. Because the orientation of the transition moments of the B_{2u}, B_{1u}, E_{1ua}, and E_{1ub} bands differs, so will the response of these moments to variations in the environment. Thus, the signals will change as the torsion angle ϕ_{CN} changes. Flexible nucleosides, like uridine (Urd) (Fig. 3.3) will have smaller CD spectra than the sterically more restricted arabinosyl analogue (Fig. 3.3). The synthetic α-analogue, in which the base is below the sugar, shows an inverted Cotton effect, while, if the base is sterically blocked in the *syn* conformation (as in m^6-Urd), small inverted Cotton effects will be seen (Fig. 3.3).

The 5'-hydroxymethylene group of the sugar can assume three staggered conformations (Fig. 3.4). They interconvert rapidly in solution, and their contributions can be evaluated from the coupling constants between $H_{4'}$, $H_{5'}$, and $H_{5''}$. Still, it must be remembered that, in nucleic acids, the conformation is *gauche-gauche*, i.e., the 5'-phosphate group is pointed toward the pentose ring.

The furanose ring also shows some flexibility of its five atoms, which can be above or below an operational mean plane (pucker, Fig. 3.5). If an atom is above the mean plane, it is called *endo*, if below, *exo*. In the absence of specific

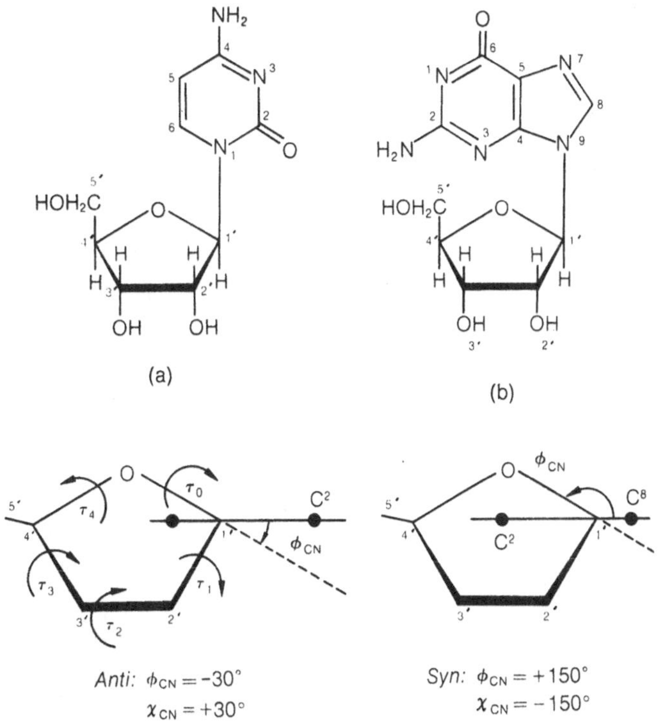

(a)

(b)

Anti: $\phi_{CN} = -30°$
$\chi_{CN} = +30°$

Syn: $\phi_{CN} = +150°$
$\chi_{CN} = -150°$

FIGURE 3.2. Cyd in the *anti* conformation (a) and Guo in the *syn* conformation (b). Definition of the glycosidic torsion angle ϕ_{CN} ($= -\chi_{CN}$) for purine nucleoside. In (c) (*anti*) are also shown the pseudorotational angles of the ribose molecule; thus τ_1 is the angle formed between the planes formed between bonds $O_{1'}$-$C_{1'}$ and $C_{2'}$-$C_{3'}$ and so on. The torsion angle for the *syn* conformation is also shown (d).

constraints, it is probable that all these conformations will occur and will rapidly equilibrate, e.g., in solution.

The furanose can be assimilated to a cyclopentane ring, and its angles are related by the so-called pseudorotation relation (4); in other words, given one angle in the ring, the other angles are determined. Two parameters will determine the conformation of the ribose: τ_m, the amplitude of the pucker, and P, the pseudorotational phase angle. Any of the five angles of the ribose will be determined by

$$\tau_i = \tau_m \cos(P + [i - 2]144)$$

Figure 3.6 shows the pseudorotation wheel with all possible puckers P of the sugar. A given value of P will automatically determine the conformational state of the sugar. Deviation from planarity will be determined by τ_m, which will differ for each nucleoside.

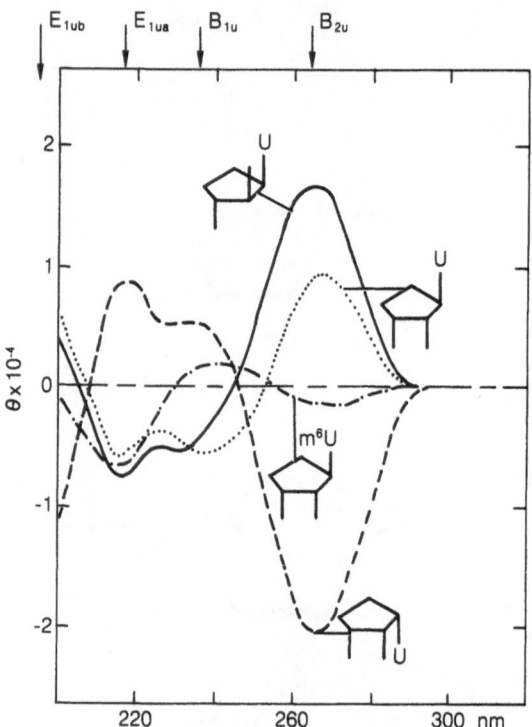

FIGURE 3.3. Circular dichroism spectra of various Urd analogues. Schematic formula show the various Urd derivatives: Urd ($\cdots$), araUrd (——), α-Urd (- - -), m⁶Urd (-·-). Approximate positions of the π-π^* transitions are indicated at the top.

FIGURE 3.4. The three possible rotamers around the 4'-5' bond of ribose in the Newman projection.

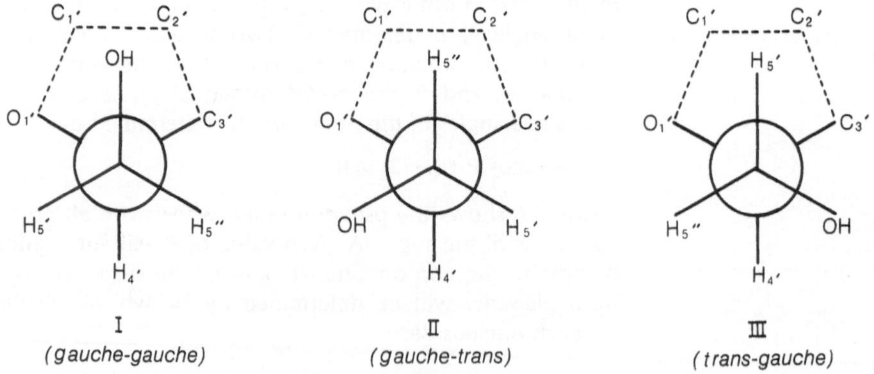

I
(gauche-gauche)

II
(gauche-trans)

III
(trans-gauche)

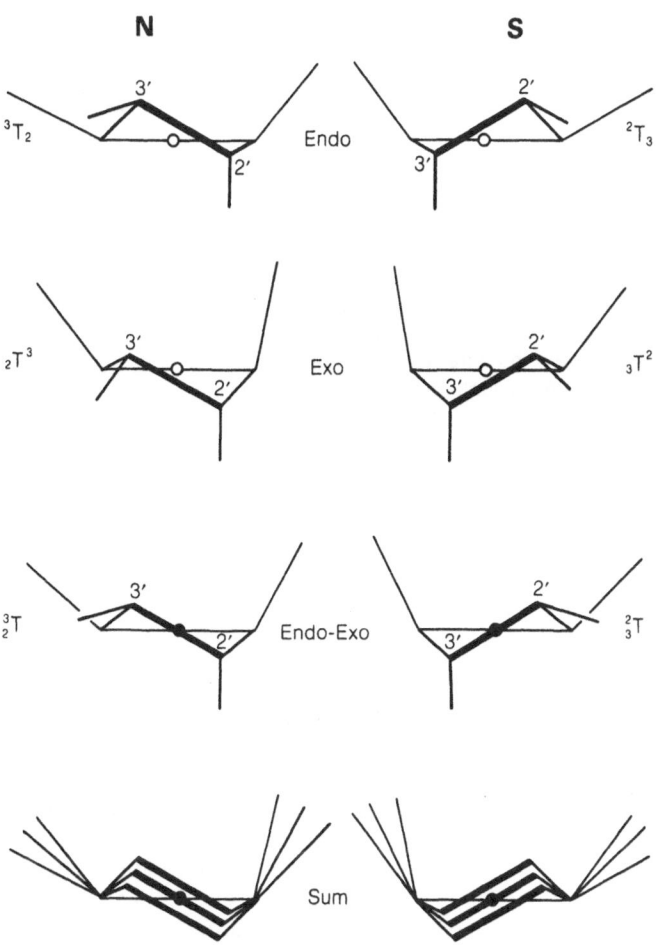

FIGURE 3.5. Main conformations of the sugar moiety of nucleosides: (N) 3'-endo-2'-exo; (S) 3'-exo-2'-endo. See also Fig. 3.6. (After Ref. 2.)

Since most nucleosides in solution will be in a flexible state, only equilibria of conformations will be observed. From empirical relations, it is now possible to determine the contributions of the two main conformation states, N(3'endo-2'exo) and S(2'endo-3'exo).

In a crystal, one, or possibly two, conformations will be favored or frozen. Two molecules of slightly different conformation in puckering, glycosidic torsion angle, and exocyclic rotamer distribution are frequently observed in the unit cell of a crystal (2,3).

Most of the single crystal studies have indicated (1-3) a predominant *anti* conformation in most natural nucleosides and nucleotides, many possible sugar puckers have been

11. Roberts, G. C. K. and Jardetzky, O. (1970). *Adv. Protein Chem.,* **24**, 447–545.

12. Jardetzky, O. and Jardetzky, C. D. (1962). *Methods Biochem. Anal.,* **9**, 235–410.

13. Bovey, F. A. (1969). *NMR Spectroscopy.* New York: Academic Press, 396 p. Becker, E. D. (1969). *High Resolution NMR.* New York: Academic Press, 310 p.

14. Abragam, A. (1961). Principles of Nuclear Magnetism. New York: Oxford University Press.

15. Schachman, H. K. (1957). *Methods Enzymol.,* **4**, 32.

16. Meselson, M., Stahl, F. M., and Vinograd, J. (1957). *Proc. Natl. Acad. Sci., U.S.A.,* **43**, 581–88.

17. Schildkraut, C., Marmur, J., and Doty, P. (1962). *J. Mol. Biol.,* **4**, 430–43.

18. Tougard, P., Chantot, J. F., and Guschlbauer, W. (1973). *Biochem. Biophys. Acta,* **308**, 9–17.

19. Wilson, H. R. (1966). *Diffraction of X-rays by Protein, Nucleic Acids, and Viruses.* London: Ed Arnold Ltd.

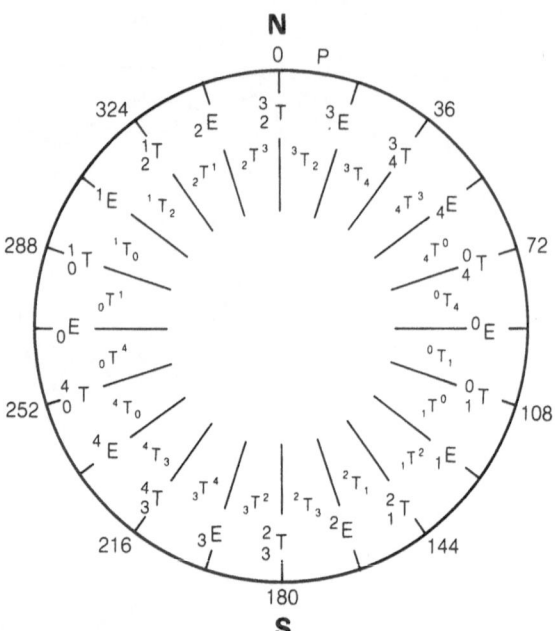

FIGURE 3.6
Pseudorotation wheel according
to Altona and Sundaralingam. The
E (envelope) and T (twist)
conformations are those where one
(E) or two (T) atoms are displaced
from the mean plane. The number
before the letter indicates the
atom with the most displacement;
the number after the letter, the
atom with the least displacement.
See also Fig. 3.5 (After Ref. 4)

Reprinted with permission from
the *J. Amer. Chem. Soc.* © 1973
by the American Chemical Society

observed with a preference of the N or S puckers, and the *gauche-gauche* rotamer conformation for the 4'-5'-bond is preferred.

On the other hand, solution studies, especially by NMR on nucleosides and nucleotides, clearly indicate that there is a multitude of conformations, which imply equilibria between the N and S sugar puckers, many ranges of ϕ_{CN}, and all rotamers around the 4'-5' bond. The concept of flexible conformations in nucleosides has become sound due to recent improvements in NMR equipment, which have permitted exact assignments and measurements of the chemical shifts of multiplets (see Chapter 2, Section 2.2) and, thus, precise determinations of coupling constants.

The multiple conformations found for sterically unhindered nucleosides are important in enzyme chemistry, where the concept of "induced fit" virtually requires flexible substrates. In contrast, one finds that sterically hindered nucleoside analogues powerfully inhibit certain enzymes. A typical case is br^8GTP, which is fixed in the *syn* conformation and which inhibits RNA-polymerase. Similar considerations could be applied to the strong cytostatic effects of arabinosides, in particular, of AraCyd (5). These compounds, in which the 2'-OH group is above the plane of the pentose (see Fig. 3.1), have a very limited rotational freedom around the glycosidic linkage. It is probably not unreasonable to believe that this lack of flexibility makes them either poor substrates for enzymes (to which they may be strongly bound

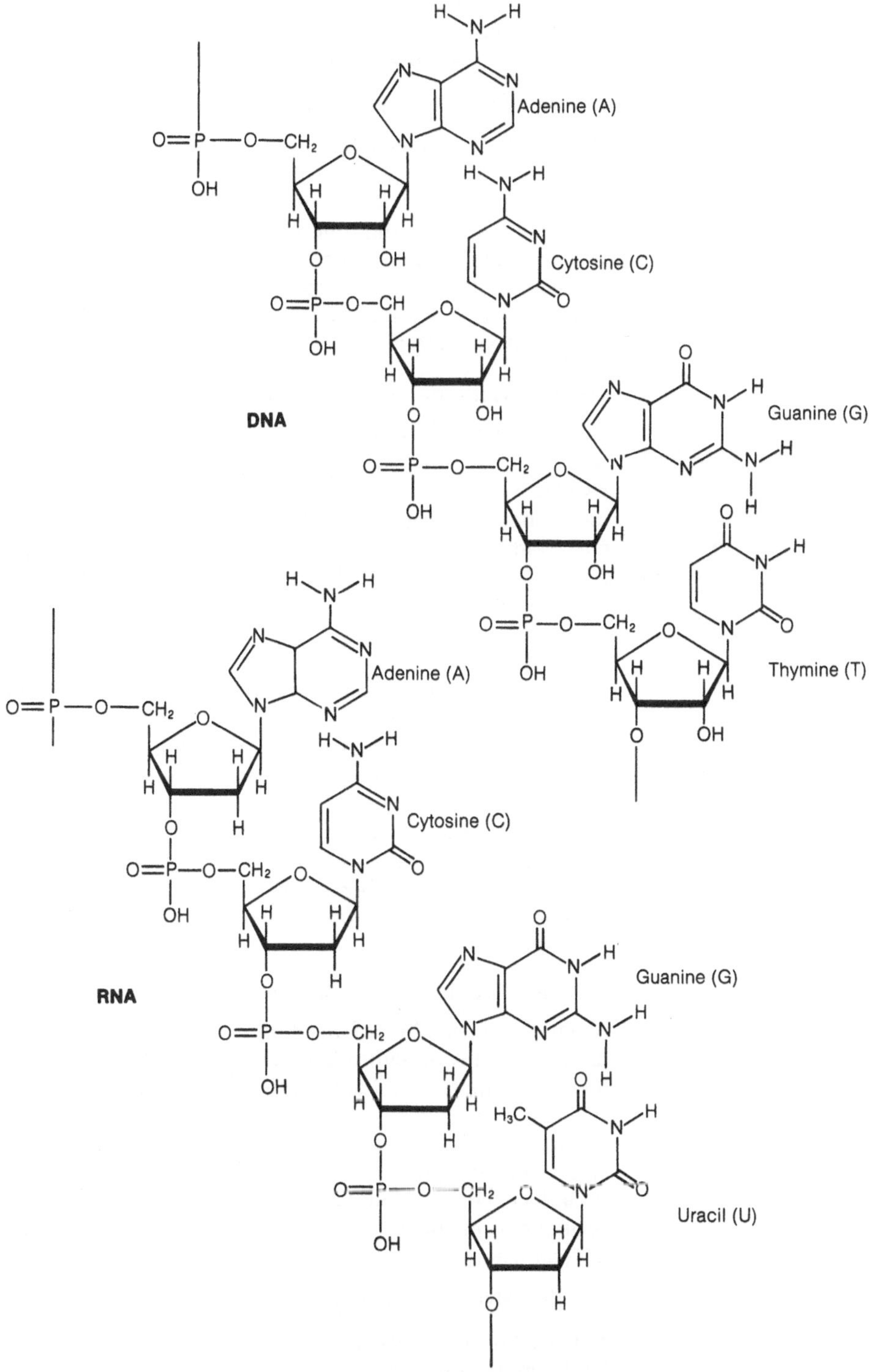

FIGURE 3.7. Phosphodiester linkage in nucleic acids.

to act as competitive inhibitors), or, if they are incorporated into the enzymatic product, they will be fixed in a rather strict conformation and thus limit function (e.g., in DNA).

3.1.4 PRIMARY STRUCTURE OF RNA AND DNA

Nucleosides are linked through a phosphodiester bond via the 3' and 5' oxygens to give the primary structure of the nucleic acids, which is ubiquitous and, without exception, the same (Fig. 3.7). The differences between DNA and RNA will be evoked in connection with the corresponding model compounds (Chapter 6, Section 6.2.3). It is essentially due to differences in the sugar pucker: DNA shows the 3'-exo conformation, RNA the 3'-endo conformation.

3.2
Chemical Reactions of Nucleic Acids and Their Constituents (6)

3.2.1 REACTIONS OF THE SUGAR MOIETY

Periodate reaction. One of the most widely employed reactions in RNA chemistry is the periodate oxidation of the cis-glycol of ribose (2' and 3'-OH). No reaction occurs with deoxyribose or arabinose. This reaction yields the terminal nucleoside as 2',3'-dialdehyde and is useful in identifying terminal nucleosides in RNA and/or coupling them with other reagents.

Color reactions. Two very old methods, which are still widely used, depend on the hydrolysis of the glycosidic linkage of purine nucleosides by hot concentrated acids, with the subsequent transformation of the sugar into furfural derivatives. These derivatives, in turn, give specific color reactions with such aldehyde specific reagents as orcinol, which reacts with furfural (from ribose) and ferric chloride to give a green color; the intensity of this color is proportional to the ribose concentration. A blue color is obtained with deoxyribose in concentrated sulfuric acid with diphenylamine (7).

3.2.2 CHEMICAL REACTIONS OF BASES

A very large number of reactions, described in the literature, imply the modification of the nucleic acid bases. A few of the most useful and widely used ones are discussed.

Nitrous acid. Sodium nitrite at low pH reacts with aromatic amines to form phenols. This reaction has been widely used to induce chemical mutations (see Chapter 7). Since the phenolic forms of the bases are not stable, the corresponding keto compounds are obtained:

Ade→Hyp, Gua→Xan, Cyt→Ura.

The reaction rate decreases in the order Cyt > Ade > Gua. Furthermore, exposed amino groups are more susceptible to attack than those involved in a structure or H-bonds.

Sodium nitrite therefore acts more efficiently on single-stranded nucleic acids than on double-stranded ones.

Dilute nitrous acid attacks Ura to yield 5-nitrouracil, which has a characteristic UV spectrum (peak at 350 nm).

Formaldehyde. Formaldehyde (like many other aldehydes or ketones) reacts with the amino groups of free bases or single-stranded nucleic acids. Since it can react with accessible amino groups, formaldehyde is used to measure the kinetics of nucleic acid unwinding (8). The chemistry is not well understood (possibly a Schiff's base as an intermediate). There is a detailed review by Feldman (9) on the subject.

Hydroxylamine. Hydroxylamine is a strong mutagen and reacts specifically with Cyt and hom^{5}Cyt (5-hydroxy-methyl-cytosine); but it also reacts less strongly with Thy and Ura. The former reaction is probably a two-step one, the first step being the addition of NH_2OH on the C^5-C^6 double bond, then the elimination of two moles of ammonia, and water, and finally, the formation of Ura.

Halogenation. Like many compounds, nucleic acid bases are attacked by halogens, the pyrimidines at position 5, the purines at position 8. Halogenobases have found wide application in chemotherapy (see Chapter 7), br^{5}dUrd is a powerful dThd analogue; br^{8}Ado and br^{8}Guo have been used in RNA polymerase studies.

Alkylation. Dimethylsulfate (or other alkylating agents) act on nucleosides and nucleic acids to yield alkylated bases, which are frequently powerful mutagens or carcinogens. The alkylation site is generally that of protonation or amino groups, i.e., N^7 (and N^1) in Gua and Ino, N^1 and N^6 in Ade, N^3 and N^4 in Cyt. Since they frequently substitute at the site where hydrogen bonds are formed, aberrant pairings are expected. Bi-functional alkylating agents, like nitrogen mustards (Cl-CH_2-CH_2-NR-CH_2-CH_2-Cl), or some antibiotics, like mitomycin, can induce cross-links between two strands of DNA by reacting with two Gua on opposite strands.

Other reactions. The reaction with P_2S_5 permits the obtention of sulfur-substituted bases, which are useful intermediates. These mercaptols (with a sulfur rather than a keto group substituent) can be treated with alcoholic ammonia to yield amines.

The literature on the chemistry of nucleosides and nucleotides is enormous. Several classics exist on this subject (e.g., Ref. 10).

Photochemistry (11). Ultraviolet irradiation of pyrimidines in aqueous solution adds a water molecule to the C^5-C^6 double bond. This reaction is reversible upon heating. If the solution is frozen, or the bases are otherwise in a rigid structure (crystals, polymers, DNA), dimers are formed between the two pyrimidines, which are thus connected through the C^5 and C^6 bonds in a cyclobutane structure, although mixed Cyt-Thy (or Cyt-Ura) dimers have also been observed. In DNA, UV irradiation forms dimers, which have a mutagenic effect. In the cell, there are specific enzymes that excise such Thy dimers and repair the break (see Chapter 4, Section 4.3.1).

3.3
Isolation of
Nucleic Acids
(12–14)

In order to study the nucleic acids by physicochemical or enzymological methods, they must be isolated and purified from other cell constituents. As their name indicates, they are acidic in character (one negative charge per nucleotide residue) and are neutralized by basic proteins (protamines, histones), polyamines (spermine, spermidine, etc.), or metallic cations (alkalis, earth alkalis). Nucleic acids are irreversibly denatured, if all their basic components are removed. For this reason, extractions are usually carried out in salt solutions buffered at pH 7.

On the other hand, DNA, being a long thin thread (20 A in diameter, several microns to millimeters or more in length), is very fragile and easily damaged by shear and mechanical forces and enzymatic activity. Great care has, therefore, to be taken to avoid breakage, if one hopes to isolate a native intact DNA. Pipetting may cause DNA breakage. Upon isolation, the DNA molecule is tested for purity and physical integrity (e.g., by sucrose density gradient centrifugation). Polysaccharides are a big problem, since they are difficult to eliminate, especially when the DNA is isolated from animal or plant tissue. The isolation procedure has to be adapted to a given purpose. Thus, it is generally not possible to isolate DNA and RNA in native form in a single procedure.

In a general procedure, after the cell wall is broken by mechanical or enzymatic methods (lysozyme), the resulting cell sap is treated with a protein-denaturing agent, such as phenol, or a detergent (dodecyl sulfate, lauryl sulfate), which precipitates proteins. Several extractions are frequently necessary. The final nucleic acid solution is treated with ethanol, to precipitate nucleic acids, or dialyzed against a suitable buffer solution. A review by Kirby (14) discusses the various isolation procedures used and their advantages and inconveniences.

3.4
Degradation and
Determination of
Nucleic Acids

3.4.1 CHEMICAL DEGRADATION

Hydrazinolysis yields apyrimidinic (i.e., DNA in which all the pyrimidines have been split at the glycosidic linkage).

Dilute acid selectively cleaves the glycosidic linkage of purines in DNA, yielding apurinic acid. It does not act on RNA, except when hot and very concentrated (see Section 3.2.1).

RNA is hydrolyzed quantitatively by 0.3 M KOH at 37°C for 24 hours, while DNA is denatured, but not hydrolyzed. This difference is the basis of the separation method of Schmidt and Thannhauser (15). The alkaline hydrolysate is treated with cold perchloric or trichloroacetic acid. The ribonucleotides remain in solution, while the denatured DNA is precipitated (with the remaining proteins) and separated by centrifugation and washing. The supernatant, containing the ribo-3'-phosphates, can now be analyzed by color reactions, chromatography, etc., while the precipitate is hydrolyzed further by acid, to be analyzed as above. This method is rather widely used for measuring the radioactivity of DNA, RNA, or proteins in *in vivo* labeling experiments using radioactive precursors. These methods give—at best—the base composition of the material under investigation, but no information about the sequence is obtained. Furthermore, due to the rather harsh treatment, losses are inevitable, and there are many sources of error.

3.4.2 ENZYMATIC DEGRADATION

There exists a number of nucleases, some of which are highly specific (Fig. 3.8). One distinguishes between endonucleases, which attack within the nucleic acid chain, and exonucleases, which attack sequentially at a given end of the polynucleotide chain.

Pancreatic ribonuclease. Ribonuclease (MW 13,600), with known amino-acid sequence and tertiary structure, is quite heat stable and frequently a great nuisance, because of its ubiquitous presence and its stability. It attacks only single-stranded RNA and cleaves at the 5'-side of pyrimidine nucleotides (see Fig. 3.8). The reaction, therefore, yields oligonucleotides with -Yp, i.e., with the 3'-phosphate group on the terminal pyrimidine nucleoside. Its reaction mechanism is well known and closely follows that of alkaline hydrolysis, i.e., the intermediate is a pyrimidine nucleoside-2',3'-cyclic phosphate.

Takadiesterases T_1 and T_2. The first of these two endonucleases splits RNA after Guo, in a way analogous to that of RNAase, and yields Guo-3'-P, while takadiesterase T_2 splits after Ado and yields Ado-3'-P.

Snake venom phosphodiesterase and polynucleotide phosphorylase. Snake venom diesterase is an exonuclease that hydrolyzes both DNA and RNA; it can be isolated from the venom of many poisonous snakes. It attacks sequentially from the 3'-OH end and yields 5'-phosphates. Polynucleotide phosphorylase (PNPase) has a similar mode of action. As the name indicates, it is a phosphorylase and not a

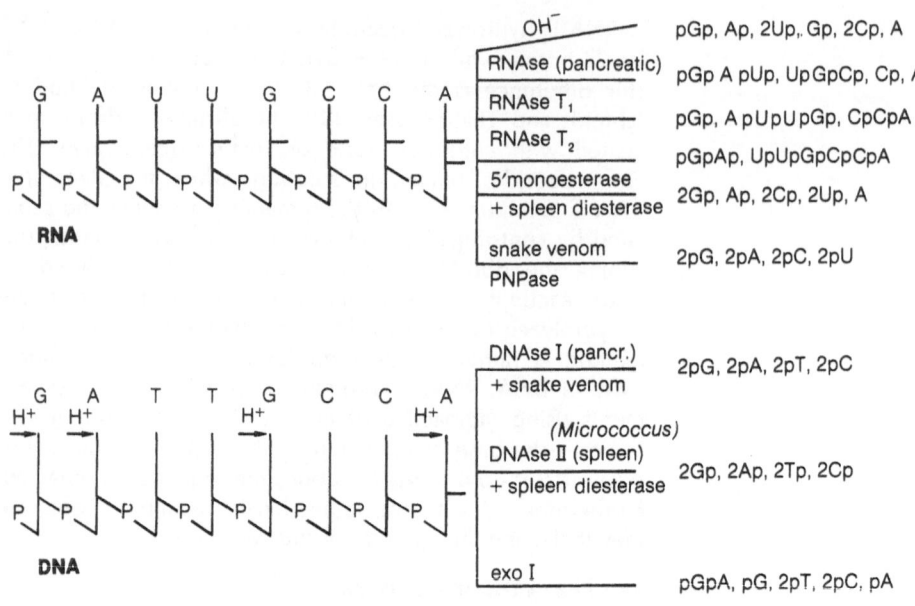

FIGURE 3.8. Chemical and enzymatic degradation of nucleic acids.

hydrolase. In the presence of excess inorganic phosphate, it degrades single-stranded RNA from the 3'-OH end to yield ribonucleoside diphosphates. It is inhibited by DNA. Note, both enzymes require a free 3'-OH terminus. If a phosphate is present on the terminal 3'-position the exonucleolytic action of the enzyme is prevented, and can continue only after the 3'-phosphate has been split off by phosphatase. Polynucleotide phosphorylase splits trinucleotides poorly and dinucleotides not at all; it stops at the tri- or di-nucleotide level. Both enzymes act sequentially, but cannot be used for sequence determinations, because they do not act in a synchronous fashion and do not follow classical Michaelis-Menten kinetics. After each reaction step, the enzyme-polymer-complex does not dissociate. The enzyme remains attached to the polynucleotide chain and continues to degrade (or to synthesize, in the case of the polymerization) very rapidly (several 1000 nucleotides per second) (16,17). This "processive" mechanism is followed by most exonucleases and polymerases.

Spleen (and micrococcal) phosphodiesterase. These exonucleases act on DNA and RNA in a direction that is the reverse of snake venom phosphodiesterase, i.e., they start from the 5'-OH and yield sequentially 3'-phosphates. They are used for nearest neighbor frequency determinations (see Chapter 4, Section 4.2.3).

Desoxyribonucleases (18,19). There are many enzymes that specifically split DNA, either in the double-stranded or single-stranded form; normally, they prefer one or the other structure.

Pancreatic DNase. Digestion of DNA by pancreatic DNase shows a considerable lag time. During this time, no apparent decrease of molecular weight of the DNA is observed. The enzyme, an endonuclease (Fig. 3.8) first splits only one chain (haplotomic break) randomly, and only when the opposite chain is broken at the same (or a nearby) position (again randomly) do smaller fragments appear. Eventually, complete digestion will occur, yielding 5'-phosphate termini.

Acid spleen DNase. This endonuclease shows no lag phase because it breaks both chains at the same point (diplotomic break) by a one-hit mechanism (18). A haplotomic mechanism also occurs. This enzyme is a dimer and splits preferentially on alternating G-C pairs, which have an internal twofold symmetry (analogous to the case of the restriction enzymes, see below). Digestion yields 3'-phosphate termini.

Escherichia coli endonuclease I. This contaminant of DNA polymerase acts by both diplotomic and haplotomic mechanisms, but yields 5'-phosphate termini.

Escherichia coli exonuclease I specifically degrades single-stranded DNA from the 3'-OH end and yields 5'-phosphates. The final dinucleotide cannot be split by the enzyme.

Escherichia coli exonuclease II is part of the DNA polymerase molecule and degrades both single- and double-stranded DNA to deoxyriboside-5'-phosphates, acting from either end of the chain.

Escherichia coli exonuclease III. This interesting enzyme has two activities: a DNA phosphatase activity, which eliminates 3'-phosphates, and an exonuclease activity, which starts from the 3'-OH group of double-stranded DNA's to yield 5'-phosphates. It will selectively digest both strands, but only from the 3'-OH end, leaving the 5'-P-strand intact. When the digestion has progressed to about 45% the enzyme stops, since sufficient double-stranded DNA is no longer present, and single-stranded DNA is not a substrate. This enzyme has been employed to obtain cohesive ends in phage DNA (see Chapter 5). Figure 3.9 illustrates this digestive mechanism.

Restriction enzymes. Restriction enzymes are a particular class of endonucleases that act on DNA and recognize and

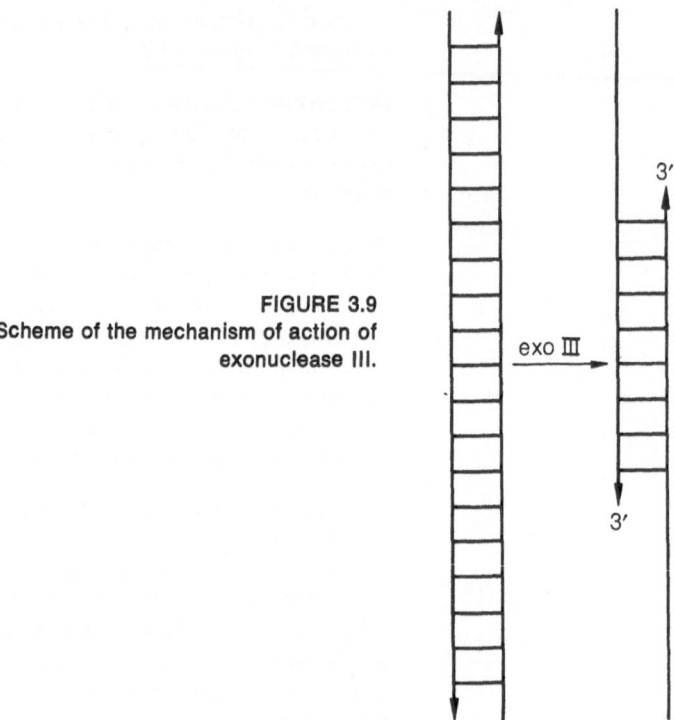

FIGURE 3.9
Scheme of the mechanism of action of
exonuclease III.

cleave very specific sequences. These sequences always have
a rotational twofold symmetry, the enzymes being dimers.
They will be treated in some detail in Chapter 5, Section
5.2.5.

References 1. Haschemeyer, A. E. V. and Rich, A. (1967). *J. Mol. Biol.*, **27**, 369–85.

2. Saenger, W. (1973). Angew. Chemie. (Int. Ed.), **12**, 591–601.

3. Voet, D. and Rich, A. (1970). *Prog. Nucleic Acid. Res. Mol. Biol.*, **10**, 183–265.

4. Altona, C. and Sundaralingam, M. (1973). *J. Amer. Chem. Soc.*, **95**, 2333–44.

5. Cohen, S. S. (1967). *Prog. Nucleic Acid Res. Mol. Biol.*, **5**, 1–62.

6. Kochetkov, N. K. and Budowsky, E. I. (1969). *Prog. Nucleic Acid Res. Mol. Biol.*, **9**, 403–38.

7. Dische, Z. (1955). *Nucleic Acids*, **1**, 285–305.

8. Lazurkin, Yu. S., Frank-Kamenetskii, M. D., and Trifonov, E. N. (1970) *Biopolymers*, **9**, 1253.

9. Feldmann, M. Ya. (1973). *Prog. Nucleic Acid Res. Mol. Biol.*, **13**, 1–50.

10. Michelson, A. M. (1963). *The Chemistry of Nucleosides and Nucleotides.* New York: Academic Press; Zorbach, W. W. and Tipson, R. S. (1968). *Synthetic Procedures in Nucleic Acid Research,* Vol. 1. New York: Wiley.

11. Lomant, A. J. and Fresco, J. R. (1972). *Prog. Nucleic Acid Res. Mol. Biol.*, **12**, 1–28.

12. Marmur, J. (1963). *J. Mol. Biol.* **3**, 208–18.

13. Cantoni, G. L., and Davies, D. R. (eds). *Procedures Nucleic Acid Research*, New York: Harper & Row, 535–614.

14. Kirby, K. S. (1964). *Prog. Nucleic Acid Res. Mol. Biol.*, **3**, 1–31.

15. Volkin, E. and Cohn, W. E. (1955). *Methods Biochem. Anal.*, **1**, 237–70.

16. Thang, M. N., Guschlbauer, W., Zachau, H. G., and Grunberg-Manago, M. (1967). *J. Mol. Biol.*, **26**, 403–30.

17. Lazarus, H. M. and Sporn, M. B. (1967). *Proc. Natl. Acad. Sci., U.S.A.*, **57**, 1386–93.

18. Bernardi, G. (1968). *Adv. Enzymol.*, **31**, 1–50.

19. Lehmann, I. R. (1963). *Prog. Nucleic Acid Res. Mol. Biol.*, **2**, 84–123.

4 Structure and Function of DNA

For DNA to fulfill its function of transmission of the genetic message, the cell must produce two identical copies of DNA during cell division, so every cell has a complete set of genetic information. The double helical structure of DNA proposed by Watson and Crick not only accounted for all the chemical and physical data then available on DNA, but also provided an understanding of the fidelity of reproduction of the DNA molecule. The model was based on the following observations:

1. Chargaff's group (1) had observed that DNA did not contain Ade, Gua, Thy, and Cyt in equal amounts, as was long believed, but that the percentage of bases in DNA from a variety of sources varied widely. But all DNA's had a regularity in common: A and T on one hand and G and C on the other hand, always occurred in equal amounts. Furthermore, the amount of A+G equaled the amount of C+T (Table 4.1).

2. If DNA was isolated carefully, a viscous solution, indicative of a macromolecule, was obtained. This macromolecule had the hydrodynamic properties of a rigid rod. When the native DNA molecule was heated or treated with alkali or acid, not only did its optical properties change, but its hydrodynamic properties as well, becoming those of a highly flexible random coil (2).

3. Finally, the X-ray fiber diagrams obtained by the Astbury group in 1938, which led to the "pile of pennies" model, encouraged a group at King's College to seek better quality diffraction patterns. The fiber diagrams of Franklin and Wilkins, obtained from native DNA, showed that the DNA molecule had considerable regularity, with a microcrystalline structure (3,4).

It should not be forgotten that two other concepts were

Table 4.1   Base Composition of DNA's from Various Sources

Source	A	G	C	T	hom⁵C	(G+C)/(A+T)
Phage T₁	26.0	24.0	24.0	26.0	0.0	0.93
Phage T₂	32.5	18.2	0.0	32.5	16.8	0.54
Mycobacterium phlei	15.8	34.5	34.2	15.5		2.17
Pseudomonas aeruginosa	16.8	33.0	34.0	16.2		2.03
Serratia marcescens	21.1	29.0	29.0	20.9		1.38
Azotobacter vinelandii	21.9	28.0	28.0	22.1		1.27
Escherichia coli K12	24.6	25.6	25.6	23.4		1.07
Bacillus subtilis	28.9	21.0	21.4	28.7		0.73
Micrococcus aureus	31.0	18.5	19.2	31.2		0.61
Clostridium perfringens	34.1	15.8	15.1	35.0		0.45
Saccharomyces cerevisiae	31.7	18.3	17.4	32.6		0.56
Calf thymus	29.2	21.9	21.9	27.1		0.78
Salmon sperm	29.7	20.8	20.4	29.1		0.70
Wheat germ	26.5	23.5	17.2	27.0	5.8	0.87

important factors in the discovery of the double helical structure of DNA: Schrödinger's (5) postulate that an "aperiodic crystal" must be the carrier of hereditary information and the work of Pauling on the structure of proteins, which had led to the discovery of the α helix (6). Crick's (7) theoretical treatment of helical structures and Pauling's triple-stranded DNA model (8) led Watson and Crick to propose the double helical structure of DNA in 1953 (9). This structure, although its derivation was highly intuitive, has proven to be essentially correct and has been modified only in detail since.

In the DNA double helix, two strands containing the bases on a deoxyribose-3'-5'-phosphodiester backbone run in opposite directions, in an antiparallel sense (a better expression is opposite polarity) (Fig. 4.1). Each base is uniquely paired with a base on the opposite strand through hydrogen bonds: only adenine (Ade) can pair with thymine (Thy) and guanine (Gua) with cytosine (Cyt) (Fig. 4.2). The two chains, therefore, have an internal complementarity, which perfectly fits the data of Chargaff. The antipolarity of the two chains has an important structural consequence: a twofold symmetry axis through the base pair, perpendicular to the helix axis, permits the exchange of complementary bases (by rotating 180° around the dyad axis) with no change in the relative positions of the attached sugar residues. In other words, if one turns the Watson–Crick helix upside down, *the two chains will look exactly alike* (see Fig. 4.3). The bases are arranged in each strand so that there is no space between two successive neighboring bases: they are stacked at a distance of 3.4 A. This distance has been obtained from X-ray diffraction patterns and is the

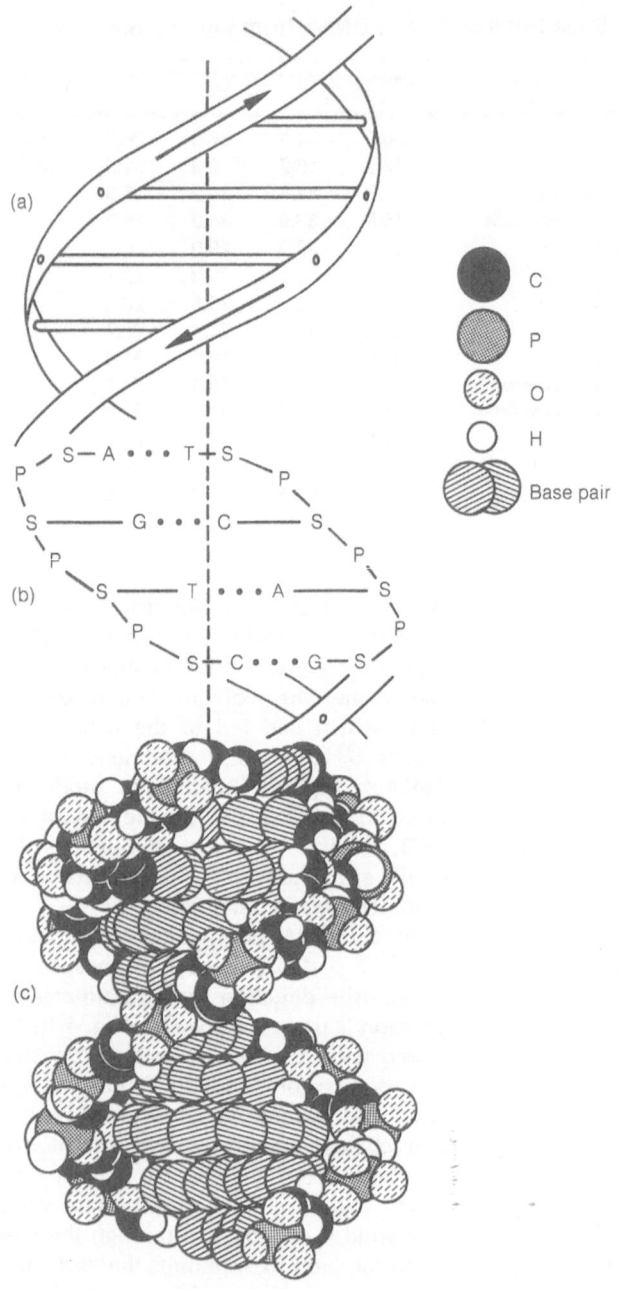

FIGURE 4.1. The helix of DNA in three presentations. (a) Ribbon ladder; (b) a somewhat more detailed letter presentation; (c) space-filling model.

From Bennett and Frieden *Modern Topics in Biochemistry.* © 1966 Macmillan Publishing Co. Inc., New York.

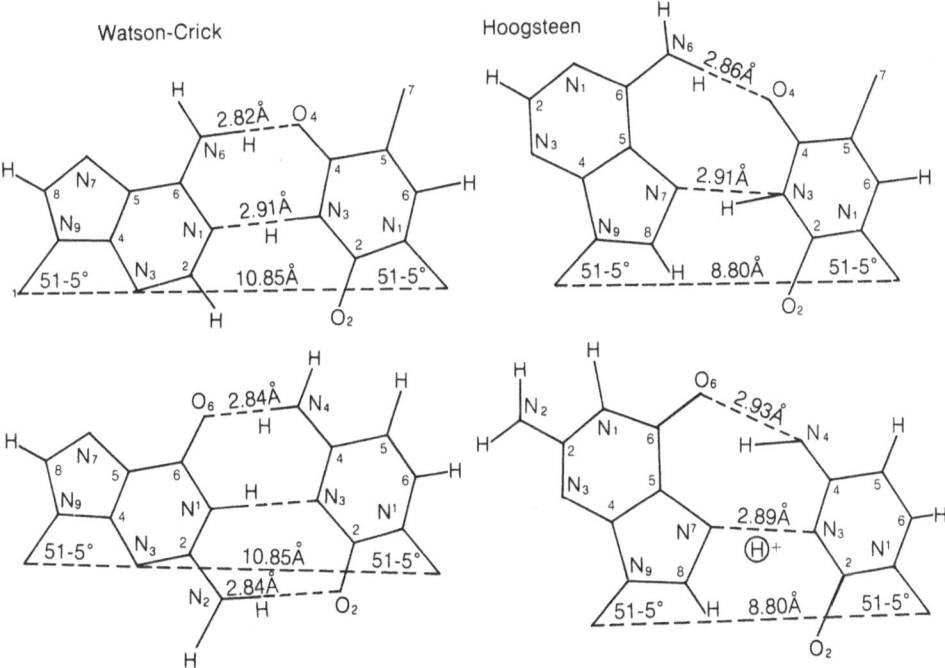

FIGURE 4.2. Coordinates of Watson–Crick and Hoogsteen pairs. (After Ref. 20.)

same as the van der Waals thickness of aromatic molecules. Another important parameter obtained from crystallography was the pitch and symmetry of the helix, which, in the B form (at high relative humidity), were found to be 34 A and 10/1 (i.e., there are ten base pairs in one turn of the helix). There are several additional features to be noted in the construction of the double helix:

1. The pairing of the bases and their stacking, at a 36° torsion angle, give rise to a small groove and a large groove.
2. The base pairs are essentially perpendicular to the helix axis (they deviate 6° from the normal of the helix axis, called the tilt).
3. The sugar phosphate backbone points toward the outside giving rise to two hydrophilic ridges with one negative charge per phosphate (cation binding), while the stacked bases form a hydrophobic core. The amino and keto groups of the bases point into the grooves and, thus, permit interactions with the solvent.
4. Stretching the helix unwinds it, and the bases unstack to a maximum distance of about 6.8 A. We shall see later that this distance is very important, since it permits the insertion of another 3.4 A thick aromatic residue between two bases (intercalation).

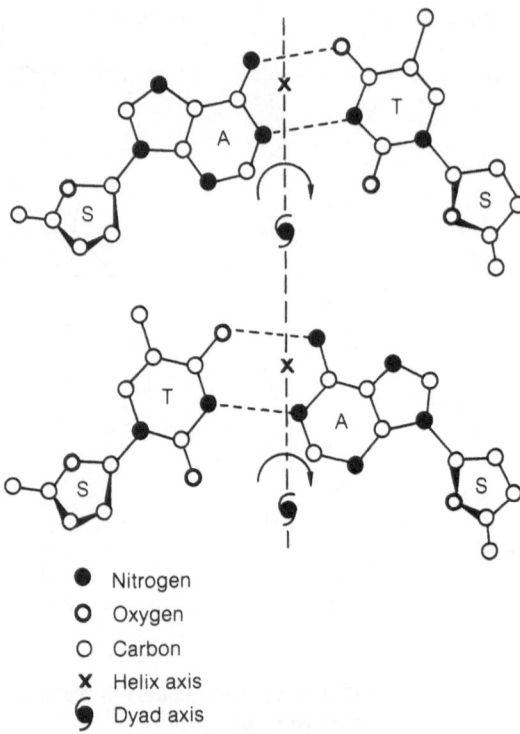

● Nitrogen
○ Oxygen
○ Carbon
✗ Helix axis
🎝 Dyad axis

FIGURE 4.3. Interrelation between an A-T pair and
a T-A pair in the Watson–Crick helix.

5. The double helix structure not only takes into account
   the equalities A＝T and G＝C, but also predicts sequence
   homologies. Thus, a sequence ApG will face CpT on the
   other strand. But ApT will be paired with ApT. This important aspect of the DNA helix is the basis of Kornberg's nearest neighbor frequency experiment, which is
   probably the most important proof of the opposite
   polarity of the two strands (Section 4.2.3).

   Watson and Crick immediately recognized the biological importance of their model and, in a second communi-
cation (10), showed that the complementarity of the two
chains provided an explanation of the duplication of genetic
material during cell division. Suppose that two strands W
and C are separated and each is used as a template for the
synthesis of a complementary strand (i.e., a strand C', complementary to W, and another, W', complementary to C).
If this were indeed the mechanism, because of base complementarity, W and W', and C and C' would have to be
the same and, thus, W-C' and W'-C would be daughter
helices, identical to each other and to the parent helix W-C.

4.2 Experimental Tests of the Watson–Crick Hypothesis

4.2.1 MESELSON AND STAHL EXPERIMENT

To prove the hypothesis of Watson and Crick, Meselson and Stahl (11) devised an experiment to distinguish between parent and progeny DNA. They used for this purpose a heavy isotope of nitrogen, ^{15}N, to label newly synthesized DNA strands.

A priori, several schemes of replication can be envisaged (Fig. 4.4). A dispersive replication, i.e., breakage and distribution of parent and daughter strands in an uneven way, seems not to be very likely in view of the constancy and sequentiality of genetic information. It can, however, not be excluded, and this hypothesis has been tested and dismissed by Rolfe (12). The conservative and semiconservative modes of replication (Fig. 4.4) differ in the way the daughter strands are distributed. The Meselson-Stahl experiment used the difference of buoyant density of DNA containing different isotopes of nitrogen. By growing *E.coli* for many generations no ^{15}N-ammonium chloride as the only nitrogen source they obtained heavy DNA, with a ^{15}N label on all nitrogen atoms of the DNA. Replication for one generation in ^{14}N-ammonium chloride yielded only DNA, intermediate in density between the heavy and normal ("light") DNA. Replication for two generations gave both light and medium density DNA (Fig. 4.5). Comparison with the schemes in Fig. 4.4 clearly shows that the semiconservative

FIGURE 4.4. Semiconservative (a) and conservative (b) replication scheme.

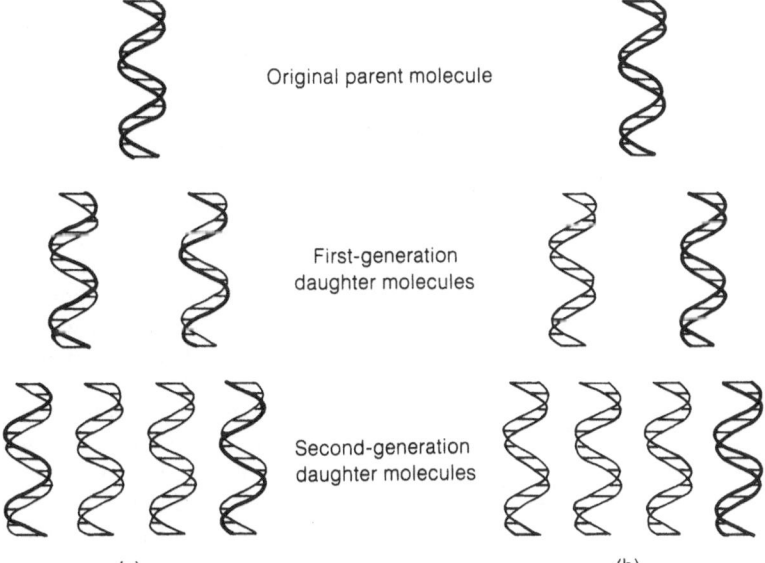

Original parent molecule

First-generation daughter molecules

Second-generation daughter molecules

(a)   (b)

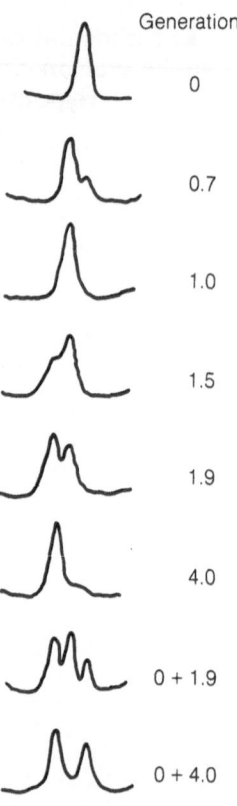

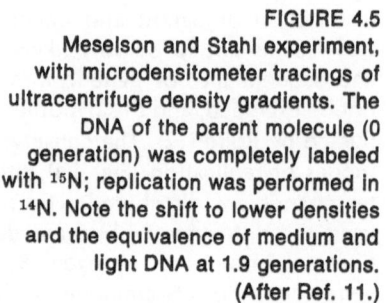

FIGURE 4.5
Meselson and Stahl experiment,
with microdensitometer tracings of
ultracentrifuge density gradients. The
DNA of the parent molecule (0
generation) was completely labeled
with ^{15}N; replication was performed in
^{14}N. Note the shift to lower densities
and the equivalence of medium and
light DNA at 1.9 generations.
(After Ref. 11.)

replication mechanism is the correct one. In contrast, the conservative mode (Fig. 4.4) would yield heavy and light material in the first generation and three times as much light as heavy DNA in the second one; no medium-density hybrid DNA should appear.

4.2.2 ROLFE'S EXPERIMENT

The Meselson–Stahl experiment *per se* did not exclude dispersive replication, although the existence of such replication was tested in a rather simple way: heat denaturation of the newly synthesized material gave two discrete peaks, one corresponding to the heavy and the other to the light strand. Rolfe (12) devised an additional experiment. By labeling with ^{13}C and ^{15}N to increase the density differences, he repeated the Meselson–Stahl experiment with one modification: he sonicated the synthesized DNA, breaking it into smaller fragments, and then performed the density gradient. The results were analogous to those of Meselson and Stahl: no all-heavy or all-light DNA could be found after one generation. The sonication procedure precluded the possibility that parts of the DNA might have been replicating in a conservative way and thus be randomly distributed among different molecules.

4.2.3 BALDWIN AND SHOOTER EXPERIMENT

Cavalieri and Rosenberg (13) suggested that a conservative replication mechanism might still occur with a "biunial" or four-stranded DNA as an intermediate. This possibility was disproved, and further confirmation of the semiconservative replication mechanism was obtained with the use of a bromodeoxyuridine label on DNA by Baldwin and Shooter (14). This thymidine analogue considerably increases the buoyant density and lowers the alkaline denaturation pK of DNA when incorporated. Hybrid DNA is formed after one generation, analogous to the Meselson-Stahl experiment. This can now be verified not only by the difference in density of the hybrid DNA, but also by titration melting experiments. Thus, at a pH intermediate between the pK's of light and heavy double-stranded DNA, hybrid DNA will dissociate into two single strand bands, characteristic of the two different components, one unlabeled, the other containing br⁵dUrd. In no case were double-stranded species containing only dThd or only br⁵dUrd observed, neither were DNA species isolated at the pK intermediate between the two DNA species observed in the double-stranded form (in the case of the biunial model), but only as single strands.

Finally, Luzzati et al. (15) showed by X-ray scattering that the mass per unit length of DNA is only consistent with a double helix structure. This result was obtained with DNA extracted from resting or growing E. coli cells. It clearly supports the semiconservative mechanism and also demonstrates the absence of biunial molecules.

4.2.4 NEAREST NEIGHBOR FREQUENCIES

When DNA is replicated by DNA polymerase (see below) a new daughter strand is synthesized from each parent strand, as was shown by Meselson and Stahl (11). Additional information on the structure of DNA has been obtained by the experiments of Josse et al. (16). Their work demonstrated the opposite polarity of the two strands. Their basic experimental design is the following: in four parallel experiments (Fig. 4.6), the same DNA is replicated with DNA polymerase, according to the following scheme:

$$n_1 \text{ dATP} + n_1 \text{ dTTP} + n_2 \text{ dGTP} + n_2 \text{ dCTP} + \text{DNA} + \text{Mg} + \text{polymerase} \rightarrow$$
$$2 (n_1 + n_2) \text{ (dAMP-dGMP-dCMP-dTMP)} - \text{DNA} + 2 (n_1 + n_2) \text{ PP}_i$$

In each experiment, one different nucleoside triphosphate

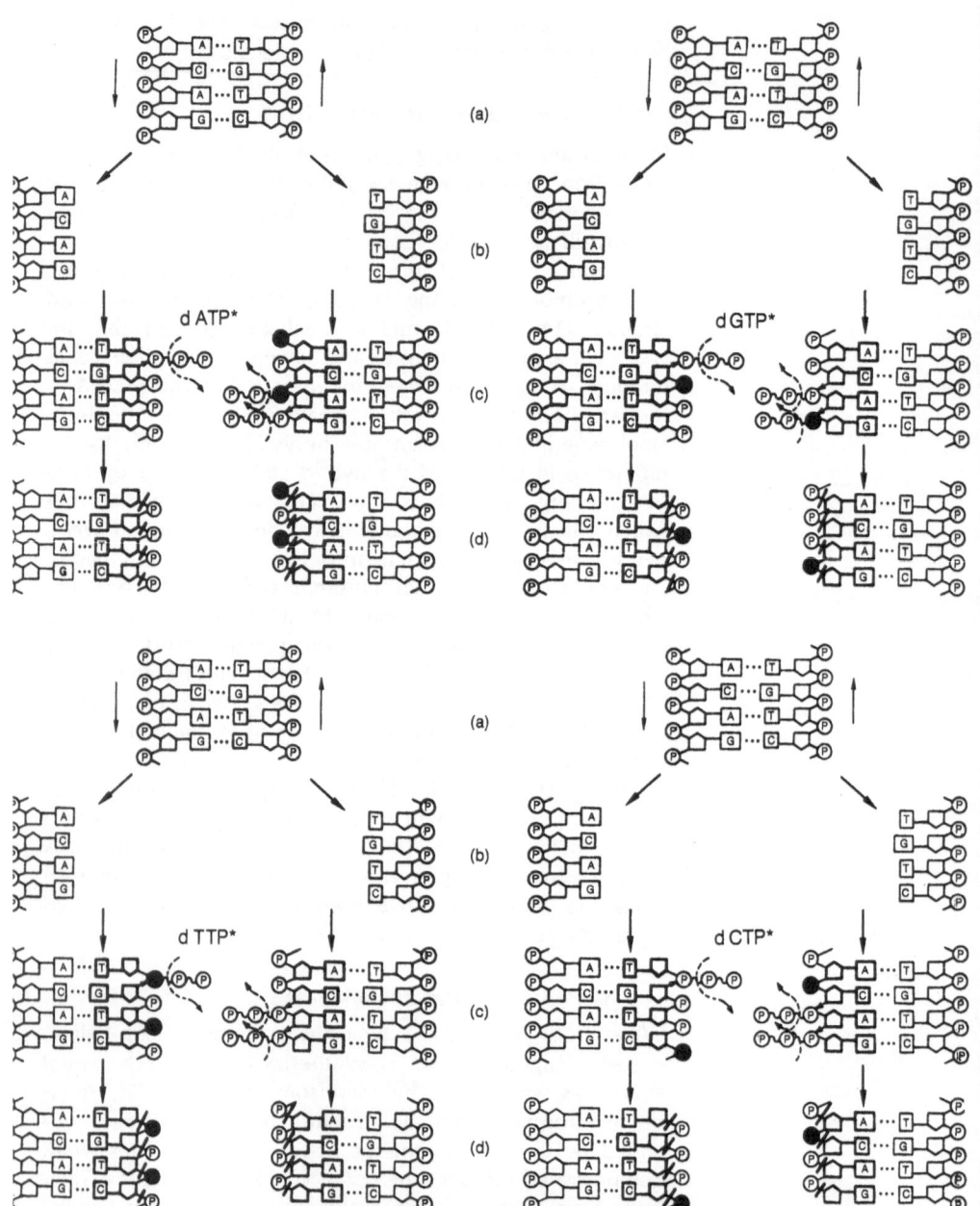

FIGURE 4.6. Nearest neighbor frequency experiment of Josse et al. (16). (a) Parent double helix; (b) Template strands; (c) Strand replication; (d) Daughter strands and their enzymatic digestion by micrococcal and spleen diesterase (/). Top: left: experiment with labeled dATP, right: with labeled dGTP; bottom: left: with labeled dTTP, right: with labeled dCTP. Labeling in replicated DNA (•).

After Bennett and Frieden Modern Topics in Biochemistry. © 1966 Macmillan Publishing Co. Inc., New York.

was labeled with radioactive phosphorus (Fig. 4.6). The phosphate in the 5′-position of the labeled nucleotide was thus radioactive. Polymerization was permitted to reach a very high level (about 20 times the original amount of DNA template). In each experiment, the DNA was isolated and digested completely with micrococcal and spleen phosphodiesterase (see Fig. 3.8), to yield deoxyribonucleoside-3′-phosphates, which were then separated by chromatography. The specific activity of each nucleotide was determined. Only those nucleotides were labeled that had the labeled nucleoside triphosphate as a neighbor.

Two kinds of information were obtained from these experiments (Tables 4.2 and 4.3).

1. Base composition: from the radioactivity measurements in Table 4.2, the relative amount of a given base recovered as 3′-nucleotide (from the DNA synthesized) must be equal to the relative amount of this base incorporated into the DNA as nucleoside triphosphate. Four equations relating the 16 fractional values and 4 base fractions can now be written. From Table 4.2 one obtains:

$$A = 0.146\ A + 0.194\ T + 0.134\ G + 0.189\ C$$
$$T = 0.075\ A + 0.157\ T + 0.187\ G + 0.182\ C$$
$$G = 0.401\ A + 0.370\ T + 0.265\ G + 0.361\ C$$
$$C = 0.378\ A + 0.279\ T + 0.414\ G + 0.268\ C$$

Thus, $A = 0.489\ C$, $T = 0.483\ C$, $G = 1.000\ C$, or $(A+T)/(G+C) = 0.48$. Since $A + T + G + C = 1$, $A = 0.164$, $T = 0.162$, $G = 0.337$, and $C = 0.337$, which agrees with the chemical determinations ($GC = 0.674$).

2. If the incorporation fractions in Table 4.2 are weighted for the compositions obtained, the nearest neighbor frequencies in Table 4.3 can be derived.
Example: $A = 0.164$

TpA: Fraction = 0.075, Frequency = $0.164 \times 0.075 = 0.012$
ApA: Fraction = 0.146, Frequency = $0.164 \times 0.146 = 0.024$
CpA: Fraction = 0.378, Frequency = $0.164 \times 0.378 = 0.063$
GpA: Fraction = 0.401, Frequency = $0.164 \times 0.401 = 0.065$

An inspection of Table 4.3 yields several facts. We find that $GpG = CpC = 0.090$. This would be expected both in parallel and anti-parallel structures. In the latter, however, GpC should face GpC, whereas it would be opposite CpG in a parallel structure. GpC is not the same as CpG, a case for opposite polarity. Similarly, $CpA = TpG = 0.063$, which agrees with a structure of opposite polarity. In a structure of the same polarity CpA and GpT should be the same (which is here the case), and TpG should equal ApC (here too). Thus, these frequencies give no result (in this case). But $ApG = CpT = 0.045$ in an antiparallel structure. For the parallel case, ApG should equal TpC (which here is much

Table 4.2 Nearest Neighbor Frequency Experiment with *M. phlei DNA*[a]

| | Labeled Triphosphate | | | | | | | | | | |
| | Reaction 1 ppp*A | | | Reaction 2 ppp*T | | | Reaction 3 ppp*G | | | Reaction 4 ppp*C | | |
	XpY	cpm	f(XpY)	XpY	cpm	f(XpY)	XpY	cpm	f(XpY)	XpY	cpm	f(XpY)
Ap	ApA	1710	0.146	ApT	2065	0.194	ApG	2500	0.134	ApC	4300	0.189
Tp	TpA	873	0.075	TpT	1665	0.157	TpG	3403	0.187	TpC	4130	0.132
Gp	GpA	4690	0.401	GpT	3945	0.370	GpG	4960	0.265	GpC	8200	0.361
Cp	CpA	4430	0.378	CpT	2980	0.279	CpG	7730	0.414	CpC	6070	0.268
Sum		11703	1.000		10655	1.000		18680	1.000		22700	1.000

[a] Data from Ref. 16.

Table 4.3 Nearest Neighbor Frequencies of *M. phlei DNA* Obtained from the Data in Table 4.2[a,b]

Roman numerals designate the frequency pairs which should be equivalent in the Watson–Crick DNA model (opposite polarity of strands); letters designate the frequency pairs, which should be equivalent in a model with the same polarity of strands.

| | Labeled Triphosphate | | | | | | | | | | | | | | | |
| | Reaction 1 ppp*A | | | | Reaction 2 ppp*T | | | | Reaction 3 ppp*G | | | | Reaction 4 ppp*C | | | |
		XpY	f(XpY)			XpY	f(XpY)			XpY	f(XpY)			XpY	f(XpY)	
Ap	B	ApA	0.024	I	A	ApT	0.031	I	F	ApG	0.045	IV	E	ApC	0.064	V
Tp	A	TpA	0.012		B	TpT	0.026		E	TpG	0.063	II	F	TpC	0.061	III
Gp	D	GpA	0.065	III	C	GpT	0.060	V	H	GpG	0.090	VI	G	GpC	0.122	
Cp	C	CpA	0.063	II	D	CpT	0.045	IV	G	CpG	0.139		H	CpC	0.090	VI
Sum			0.164				0.162				0.337				0.337	

[a] Data from Ref. 16.

[b] The chemical analysis of *M. phlei DNA* is A: 0.162, T: 0.165, G: 0.338, C: 0.335

larger), and CpT should equal with GpA (which here too is much too large). In turn, TpC=GpA=0.063, which agrees with the parallel case. This kind of analysis has been performed for many DNA's of viral, bacterial, and animal origin (17,18) and all have demonstrated antiparallel polarity.

FIGURE 4.7. X-ray fiber difraction patterns of calf thymus DNA. (a) A-DNA, 75% relative humidity; fiber tilted 73° to X-ray beam; (b) B-DNA, 66% relative humidity; fiber tilted 13° to X-ray beam. (From Ref. 18.) (c) and (d) Projections perpendicular and along the helix axis of the two forms. Note the helix axis (+) is between the base pairs in the B form but far before them in the large groove in the A form. (From Ref. 20.)

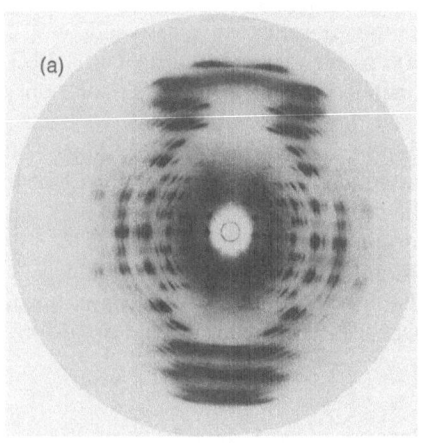
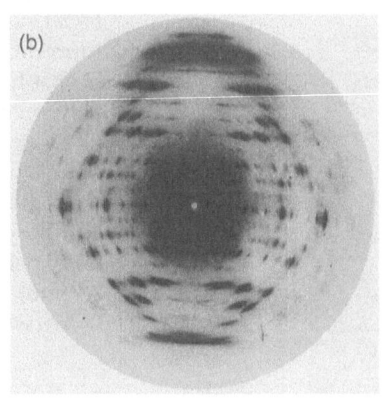

A-form   B-form

4.2.5 REFINED X-RAY ANALYSIS

The King's College group (London) has refined X-ray fiber diffraction to a great degree, and the patterns obtained (Fig. 4.7) are of such quality that many ambiguities can be eliminated. From these patterns, it is clear (despite the loss of contrast due to reproduction of their figures) that the B form is a tenfold helix and the A form an 11-fold helix. Less evident are the finer details. Arnott *et al.* (19,20) did, however, clearly show that only a Watson–Crick pair could account for the observed patterns, thus eliminating the Hoogsteen pair (see Section 5.5). But even these detailed diagrams cannot detect local, microscopic changes in the DNA structure. We shall see in Chapters 5 and 6 that this point may be of importance.

In general, DNA fibers (19) at high humidity (above 66%), and in the presence of excess salt, will give the B form diffraction pattern (tenfold helix) [Fig. 4.7(b)]. It is very probably the native DNA form found in solution and, probably, in the cell. It is characterized by base pairs that are virtually perpendicular to the helix axis [Fig. 4.8(b)]. If the fiber is less hydrated, and if the salt content of the fiber is decreased, the A form is observed [Fig. 4.7(a)], this is an 11-fold double helix, with a pitch of 28.15 A, in which the base pairs are tilted by 20° from the helix axis (Fig. 4.8, A-form). If the hydration of the B form is brought below 66%, with excess salt still present, a third form, C, is ob-

FIGURE 4.8. Schemes of the A and B forms of DNA.

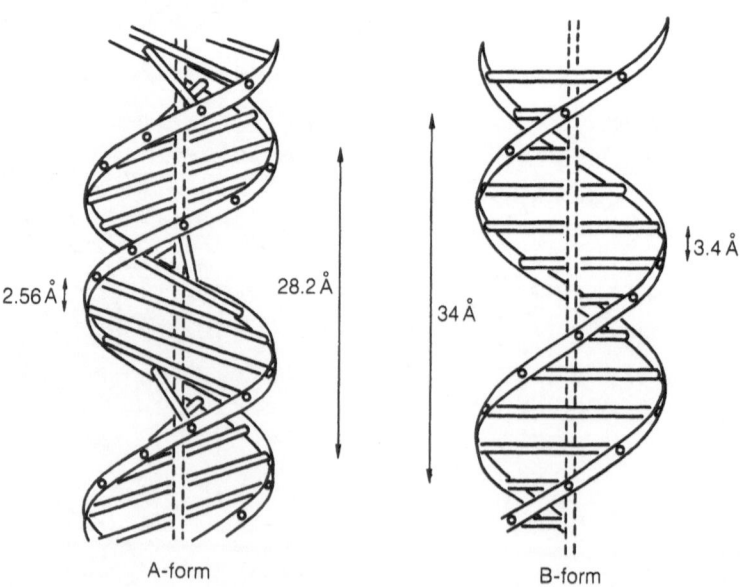

2.56 Å   28.2 Å   34 Å   3.4 Å

A-form   B-form

served which has a tilt in the opposite direction and has a 28/3 helix symmetry. In the A form the displacement of the base pairs away from the helix axis increases the depth of the large groove, but reduces its width, with a tightening of the helix. The projection of the base pairs on the helix axis (the axial rise h) is only 2.56 A in the A form, compared with 3.37 A in the B form. The most important conformational difference between A and B forms, however, is the change in sugar conformation: while the A form has 3'-*endo* puckering, the B form has the 3'-*exo* conformation (20). This difference in sugar conformation seems to be the unique structural difference between RNA (which is always and exclusively in the A form) and DNA (which can show the B→A transition) (20). Interestingly enough, RNA-DNA hybrids (21) are in the A form. It is this structural feature which supports the belief that, during transcription, the DNA molecule must undergo a B→A transition (22).

4.3
The Role of DNA in the Cellular Machinery— The Central Dogma

It is naturally quite impossible to treat the processes of replication, transcription, and translation in a comprehensive manner—particularly since these fields are constantly evolving. Thousands of articles are published every year on the subject, and numerous symposia and review articles have been devoted to these questions (23–34).

4.3.1 REPLICATION

In 1958, Kornberg et al. (35) discovered the first enzyme capable of polymerizing deoxyribonucleoside triphosphates into DNA. They called this enzyme, isolated from *E. coli,* DNA polymerase or duplicase. Three distinct DNA polymerases (*I, II,* and *III*) have since been isolated from bacterial cells and two from animal tissues (α and β). All these enzymes catalyze the following reaction:

$$n_1 \text{ dATP} + n_1 \text{ dTTP} + n_2 \text{ dGTP} + n_2 \text{ dCTP} + \text{DNA} + \text{Mg} + \text{polymerase} \rightarrow$$
$$2 (n_1 + n_2) \text{ (dAMP-dGMP-dCMP-dTMP)} - \text{DNA} + 2 (n_1 + n_2) \text{ PP}_i$$

The principal characteristics of this reaction are:

1. The reaction depends on the presence of a DNA template. RNA cannot replace DNA.
2. Initiation of the reaction is absolutely dependent on the presence of a primer with free 3'-OH groups (which can be RNA).
3. Only deoxyribonucleoside triphosphates can serve as substrates. If one of the four nucleotides is missing, the reaction rate approaches zero.
4. Magnesium ions are necessary.
5. The reaction product is DNA (up to several dozen times the initial template concentration).

Meselson and Stahl (11) found that the *in vivo* replication of the bacterial chromosome followed a semiconservative mechanism. Certain details of replication were elucidated by Cairns (24). He succeeded in obtaining radio-autographs (after incorporation of tritium-labeled dThd) with a resolution power of 1μ (Fig. 4.9). From these and other experiments the following picture emerges:

1. Replication of the bacterial chromosome (*E.coli, B. Sub-tilis*) is sequential and starts at one (or two) initiation points. The genetic data also suggest two points of initiation for DNA synthesis.
2. The replication mechanism seems to imply the existence

FIGURE 4.9. Autoradiograph of duplicating *E. coli* chromosome after two generations of tritiom labeling. The insert shows the interpretation of the autoradiograph based on the varying density of the grains. Dense segments represent doubly labeled DNA helices (B), faint segments singly labeled DNA (A). The parent chromosome began replication at X, and newly labeled strands have been synthesized to point Y. (From Ref. 24.)

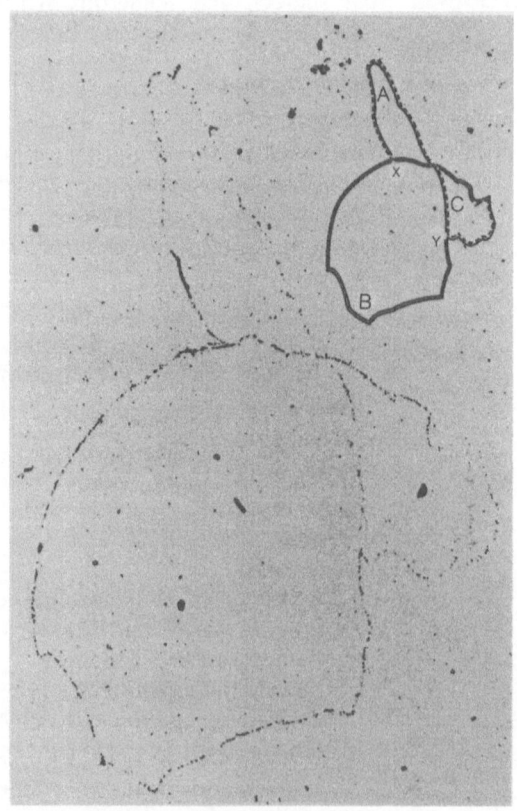

of a replication "fork," i.e., both complementary strands are synthesized simultaneously at the growing point.

3. Synthesis always takes place in the sense 5′→3′. Wake (27) observed the replication fork in replicating *B. subtilis* chromosomes by electron microscopy and showed that it proceeds in both directions at about the same rate.

4. The entire *E.coli* chromosome is a single circular structure of about 1100 μ. This circularity seems to be maintained during replication by a swivel action at the growing replication fork.

5. Okazaki et al. (25) showed that the DNA synthesized at the fork is in fragments, which are then sealed by a ligase. By using ligase-thermosensitive mutants, Newman and Hanawalt (26) confirmed this observation. At temperatures where the ligase is inactive, small DNA fragments accumulate. The Okazaki-fragments are probably initiated on an RNA primer, which is removed later.

6. The velocity of linear synthesis is about 3000 nucleotides/second, which corresponds to a speed of unwinding of the parental DNA of about 18,000 rpm.

DNA ligases have been isolated from many sources, particularly from *E.coli* during phage infection (28). These enzymes can close breaks (not gaps) in the DNA phosphodiester chain. These enzymes can also transform an open circular DNA molecule into a closed circle by a covalent 3′-5′ phosphodiester bond. Similarly, many cells possess "repair" enzymes that can repair gaps and defects in the DNA induced by various agents. Thus, Thy dimers formed by UV irradiation of DNA are hydrolyzed by specific DNases and are replaced by correct sequences by the repair enzymes. It is probable that this is the main role of the DNA polymerase of Kornberg.

The most spectacular result of Kornberg's group (23) was the *de novo* synthesis *in vitro* of a single-stranded DNA of phage φX-174 or M-13, which was infectious, by a highly purified DNA polymerase.

The vegetative form of the small *E. coli* phage φX-174 is a circular single-stranded DNA (23). This (+) strand can be copied by *E.coli* DNA polymerase to give a (−) strand. This (−) strand is maintained on the (+) strand by base pairing and can be closed by the ligase. Data from electron microscopy studies supported this synthetic double-stranded, circular, closed DNA as being the same as the replicative circular form of the phage DNA. Furthermore, all the other physicochemical properties of the two DNA's were identical (e.g., sedimentation constant). If, now, during the synthesis of the (−) strand, dTTP is substituted by brdUTP one can form a "heavy" (−) strand, which can be isolated in a density gradient. The circular form of the (−) strand was found to be infective for *E.coli*, serving as a new template for a second replication cycle to form a complementary (+) strand. One thus obtains a completely synthetic replicative form of the phage ϕX-174. At least ten different proteins are necessary for the *in vitro* reconstitution of the complete system (36).

The details of the replication mechanism are far from clear at the present time. The biggest problem is quite clearly the antipolarity of the two DNA chains, which seems not to support the $5' \rightarrow 3'$ replication mechanism of DNA polymerase. A number of models have been devised that reconcile some contradictions (25, 29–34). A lucid discussion of the state of the art is found in Refs. 31–34.

4.3.2 TRANSCRIPTION

The second process in which DNA is directly involved is the transcription of information from DNA to RNA. Some of this is messenger RNA (mRNA), which then serves as a template for the synthesis of proteins in conjunction with ribosomes and transfer RNA (tRNA), which are themselves synthesized on the DNA template. The complementary Watson–Crick base pairing mechanism is effective here as well.

RNA polymerase, an enzyme with some similarities to DNA polymerase, requires the four ribonucleoside triphosphates, a DNA template, Mg, and low concentrations (<0.1 M) of KCl. The structure of the enzyme has been studied in recent years. The so-called core-enzyme contains four subunits, two identical α chains (MW 39,000), and two different β chains (MW 155,000 and 165,000). Furthermore, several "factors" are necessary for the proper function of the enzyme; sigma (MW 90,000) is necessary for the correct initiation of the transcription, and rho (subunit MW 55,000) recognizes the termination signals on the DNA template. In the absence of sigma, nonspecific initiation is observed; in the absence of rho, very long RNA chains are found, the synthesis apparently continuing to the end of the DNA template. There are other factors that also appear to control the activity of RNA polymerase.

In contrast to replication, in which both strands are copied, transcription is asymmetric and uses only one specific strand as a template, while the other strand seems to be "silent." Transcription starts at specific initiation sites (promotors) and terminates at specific sites. Numerous recent reviews (37–39) describe the current state of the art for these mechanisms.

In viruses and phages, the transcription of specific functions (genes) can clearly be distinguished *in vivo* on the basis of their time of appearance after viral infection. They are called "early" and "late" functions, recognizable by the appearance of various enzyme activities. Regulation of transcription is accomplished by a complex system of repressors and activators, acting on the nucleotide regions immediately preceding the promotor regions where RNA polymerase is bound and imitates transcription.

Finally, newly synthesized RNA, including mRNA, ribosomal RNA (rRNA), and tRNA enter the translation cycle to form proteins. The biosynthesis of proteins is again dominated by the same Watson–Crick base pairing responsible for the specificity of DNA and its replication and transcription. Here the adaptors, tRNA, are bound to the mRNA through three base pairs and regulated by the genetic code. This system of transfer of the genetic information,

DNA→RNA→Proteins

widely called the central dogma of molecular biology, is essentially unidirectional and irreversible, at least in procaryotes (i.e., bacteria). It is directed by the same specific interactions between templates and small- or medium-sized substrates, involving small forces (hydrogen bonds), as in the DNA molecule itself. It is only very recently (40) that reverse transcriptases have been observed in tumor (and developing?) eucaryotic cells, which use RNA templates to synthesize DNA.

The problem of recognition apparently is important both in replication and transcription, although it is rarely discussed in physical terms. The question why polymerases start and stop their action at a given point (to one nucleotide exactly!) is completely open. Recent progress in the elucidation of the sequences of RNA polymerase promotor sites shows some degree of homology between these promotors (41). There must be initiation and termination signals for all these mechanisms, as well as for their control (operon, repressor sites, etc.) (42).

These findings on DNA structure, which have been described in the preceding sections, give a unified picture of one of the central problems of life. The attractive simplicity and beauty of the double helix model has exerted its fascination now for 22 years, and no data that disprove it have been forthcoming, although attempts have not been lacking. The Watson–Crick structure, and the involvement of the base pairing principle in the cellular machinery at all levels, is central to molecular biology. An important experimental demonstration has been the complete synthesis of the DNA sequence that codes for alanyl-tRNA (43).

References   1. Chargaff, E. (1960). *Experientia*, **6**, 201–9.

2. Sadron, C. (1960). *Nucleic Acids*, **3**, 1–25; Rice, S., and Doty, P. (1957). *J. Amer. Chem. Soc.*, **79**, 3937.

3. Franklin, R. E. and Gosling, R. G. (1953). *Nature*, **171**, 740–41.

4. Wilkins, M. H. F., Stokes, A. R., and Wilson, H. R. (1953). *Nature*, **171**, 738–40.

5. Schrödinger, E. (1945). *What is Life?* Cambridge: Cambridge University Press.

6. Pauling, L., Corey, R. B., and Bramson, H. R. (1951). *Proc. Nat'l Acad. Sci., U.S.A.*, **37**, 205; 729.

7. Cochrane, W., Crick, F. H. C., and Vand, V. (1952). *Acta Crystallog.*, **5**, 581; Klug, A., Crick, F. H. C., and Wyckhoff, H. S. (1958). *Acta Crystallog.*, **11**, 199.

8. Pauling, L. and Corey, R. B. (1953). *Nature*, **171**, 346–48.

9. Watson, J. D. and Crick, F. H. C. (1953). *Nature* **171**, 737–38.

10. Watson, J. D. and Crick, F. H. C. (1953). *Nature*, **171**, 964–67.

11. Meselson, M. and Stahl, F. N. (1958). *Proc. Nat'l. Acad. Sci., U.S.A.*, **44**, 671–78.

12. Rolfe, R. (1962). *J. Mol. Biol.*, **4**, 22–30.

13. Cavalieri, L. F. and Rosenberg, B. H. (1961). *Biophys. J.*, **1**, 301–51.

14. Baldwin, R. L. and Shooter, E. M. (1963). *J. Mol. Biol.*, **7**, 511–26.

15. Luzzati, V., Luzzati, D. and Masson, F. (1962). *J. Mol. Biol.*, **5**, 375–83.

16. Josse, J., Kaiser, A. D. and Kornberg, A. (1961). *J. Biol. Chem.*, **236**, 864–75.

17. Swartz, M. N., Trautner, T. A. and Kornberg, A. (1962). *J. Biol. Chem.*, **237**, 1961–67.

18. Subak-Sharpe, H. *et al.* (1966). *Cold Spring Harb. Symp. Quant. Biol.*, **31**, 737–48.

19. Arnott, S., Wilkins, M. H. F., Hamilton, L. D. and Langridge, R. (1965). *J. Mol. Biol.*, **11**, 391.

20. Arnott, S. (1970). *Prog. Biophys.* **21**, 265–319.

21. Milman, G., Chamberlin, M. and Langridge, R. (1967). *Proc. Nat'l. Acad. Sci., U.S.A.*, **57**, 1804–11.

22. Arnott, S., Fuller, W., Hodgson, A., and Prutton, I. (1968). *Nature*, **220**, 561–65.

23. Goulian, M., Kornberg, A. and Sinsheimer, R. (1967). *Proc. Nat'l. Acad. Sci., U.S.A.*, **58**, 2121–2328.

24. Cairns, A. (1961). *J. Mol. Biol.*, **3**, 756–61; (1963). *J. Mol. Biol.* **6**, 208–13.

25. Okazaki, R. *et al.* (1968). *Cold Spring Harb. Symp. Quant. Biol.*, **33**, 129–43.

26. Newmann, J. and Hanawalt, P. (1968). *Cold Spring Harb. Symp. Quant. Biol.*, **33**, 144–50.

27. Wake, R.G. (1973). *J. Mol. Biol.*, **77**, 569–75.

28. Lehmann, I. R. (1974). *Science*, **186**, 790–97.

29. Eisen, H., Pereira da Silva, L. and Jacob, F. (1968). *Cold Spring Harb. Symp. Quant. Biol.*, **33**, 755–64.

30. Gilbert, W. and Dressler, D. (1968). *Cold Spring Harb. Symp. Quant. Biol.*, **33**, 473–85.

31. Becker, A. and Hurwitz, J. (1971). *Prog. Nucleic Acid Res. Mol. Biol.*, **11**, 423–60.

32. Shekman, R., Weiner, A. and Kornberg, A. (1974). *Science* **186**,

5 Physical Chemistry of DNA– The Problems of DNA Research

5.1 Established Facts

5.1.1 MELTING OF DNA (T_m)

The UV absorption spectrum of native DNA is hypochromic compared with that of its constituents, the nucleotides. If DNA is heated to 100°C, UV absorption increases and approaches that of the monomers (Fig. 5.1). This increase is not gradual, but takes place at a given temperature (Fig. 5.2), depending on the DNA and the experimental conditions. Similarly, other properties, like viscosity, optical rotation, polarographic wave, etc., undergo sudden changes at the same temperature. This transition temperature can be compared to the melting of a crystal and has therefore also been called the "melting temperature" (T_m). It is probably the most widely used physical parameter in nucleic acid and polynucleotide research because of its high reproducibility.

Marmur and Doty (1) first observed that T_m depended on two variables: (a) the G-C content of the DNA and (b) the ionic strength of the solution. More generally, T_m depends on the solvent conditions of the polynucleotide. Therefore any deviation from aqueous solution at pH 7 has to be taken into account, e.g., addition of organic solvents and changes in salt concentration or pH. A linear relationship between T_m and G-C content has been found. At pH 7 in 0.165 M NaCl

$$T_m = 69.3 + 0.41 \, (\% \, GC)$$

It is therefore sufficient to measure T_m by a suitable technique in 0.165M NaCl to determine the G-C content (Fig. 5.3). This method is now routinely used to identify unknown DNA samples and even to determine taxonomic relations among species. There are, however, exceptions to this relationship due to anomalies of the DNA, like modifi-

987–93; Kornberg, A. (1974) *DNA synthesis*, San Francisco: Freemann.

33. Goulian, M. (1972). *Prog. Nucleic Acid Res. Mol. Biol.,* **12**, 29–48.

34. Gefter, M. L. (1974). *Prog. Nucleic Acid Res. Mol. Biol.,* **14**, 101–15.

35. Kornberg, A., Lehmann, I. R., Bessman, M. J. and Simms, E. S. (1956). *Biochim. Biophys. Acta,* **21**, 294–95.

36. Shekman, R., Weiner, T. H., Weiner, A., and Kornberg A. (1975). *J. Biol. Chem.,* **250**, 5859–65.

37. (1970). *Cold Spring Harb. Symp. Quant. Biol.,* **35**.

38. Bautz, E. K. F. (1972). *Prog. Nucleic Acid Res. Mol. Biol.,* **12**, 129–60.

39. Richardson, J. P. (1969). *Prog. Nucleic Acid Res. Mol. Biol.,* **9**, 75–116.

40. Temin, H. and Baltimore, D. (1972). *Adv. Virus Res.,* **17**, 129.

41. Pribnow, D. (1975). *Proc. Nat'l. Acad. Sci., U.S.A.,* **72**, 784–8.

42. Von Hippel, P. and McGhee, J. D. (1972). *Ann. Rev. Biochem.,* **41**, 231–300.

43. Gupta, N. K. et al. (1968). *Proc. Nat'l. Acad. Sci., U.S.A.,* **60**, 1338–43

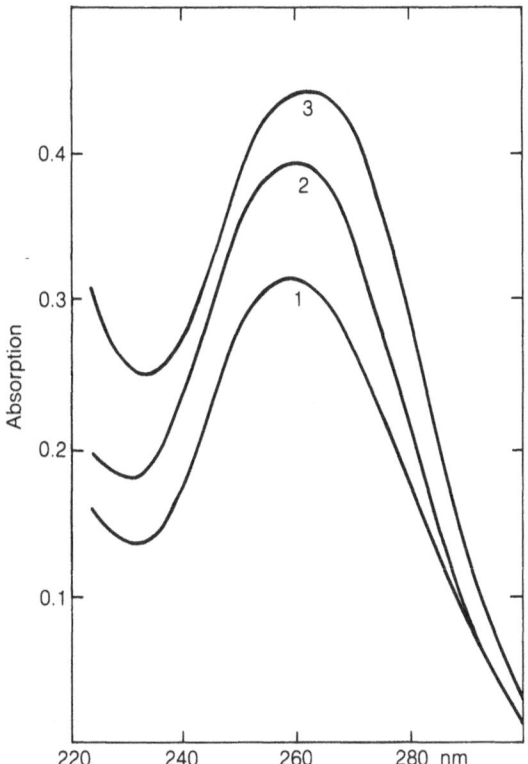

FIGURE 5.1. UV absorption spectrum of DNA in the native (1) and denatured (2) state and the sum of the nucleotides (3).

cation of certain bases (glucosylation of hom^{5}Cyt in certain phage DNA's).

Figure 5.4 shows another linear relationship of T_m, that obtained as a function of the logarithm of the cation concentration. It is followed by all DNA's and most double-stranded anti-parallel polynucleotides (see Chapter 6).

$$\Delta T_m / \Delta \log [NaCl] = 17 \text{ to } 22°$$

5.1.2 BUOYANT DENSITY VS G-C CONTENT

Schildkraut et al. (2) demonstrated another relationship between the G-C content and the buoyant density (in CsCl gradient) of a DNA sample (Fig. 5.5).

$$\rho_B = 1.66 + 0.098 \ (\% \ GC)$$

Another, nondestructive method is optical rotation: the specific rotation at 290 nm is a function of G-C content (3). A similar relationship is observed for CD at 275 nm (4).

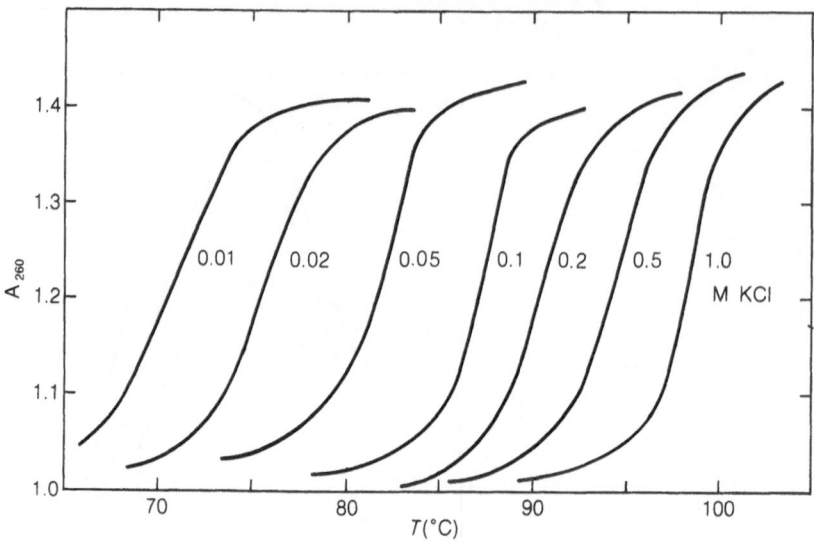

FIGURE 5.2. Melting curves of *E. coli* DNA at different KCl concentrations (1).

5.1.3 HYDRODYNAMIC PROPERTIES OF DNA

Other methods that provide quite different information on the macromolecular structure of DNA are sedimentation measurements and light scattering. The theory of sedimentation and light scattering of very large macromolecules is very complex and will not be treated here. It will suffice to point out the work of Rubenstein et al. (5) and Marmur et al. (6), which correlates sedimentation coefficient and molecular weight of DNA (Fig. 5.6). A commonly used method is to run a DNA of known molecular weight as a standard together with the sample.

Shearing must be avoided so that unbroken DNA of high molecular weight is obtained. The use of the usual bulb viscometers will break very long DNA molecules. Preparation of DNA by lysis of nuclei or chromosomes directly in the ultracentrifuge cell or in shear-free viscometers will eliminate handling of free DNA. This latter technique has been considerably advanced by Klotz and Zimm (7), who constructed a viscometer that permits the determination of very high molecular weights (of several billion Daltons). The DNA solution is placed between two concentric cylinders, one of which is fixed. The angular displacement of the other cylinder can be measured. The relaxation of the torsion applied is related to the molecular weight of the largest molecules in the solution, while the smaller molecules contribute very little to the total viscosity.

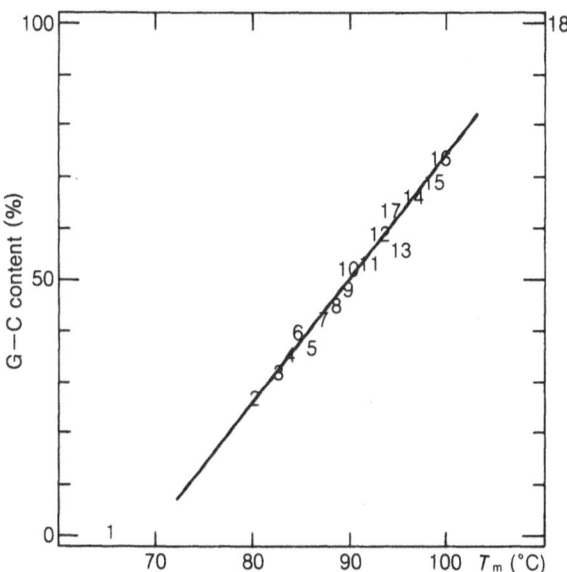

FIGURE 5.3. Dependence of T_m on G-C content. The DNA's shown are: (1) poly[(d(A-T)]; (2) *Clostridium perfringens*; (3) *Bacillus cereaus*; (4) *B. thurigiensis*, (5) *Proteus vulgaris*; (6) phage T_6; (7) *B.subtilis*; (8) *Vibrio cholerae*; (9) phage T_7; (10) *E. coli*; (11) *Neisseria meningitidis*; (12) *Aerobacter aerogenes*; (13) *Azotobacter vinelandii*; (14) *Mycobacterium phlei*; (15) *Sarcina lutea*; (16) *Micrococcus lysodeikticus*; (17) *Pseudomonas aeruginosa*; (18) poly(dG)·poly(dC). (All data from Ref. 1.)

FIGURE 5.4. Ionic strength dependence (in KCl) of *E. coli* and *Diplococcus pneumoniae* DNA. (From Ref. 1.)

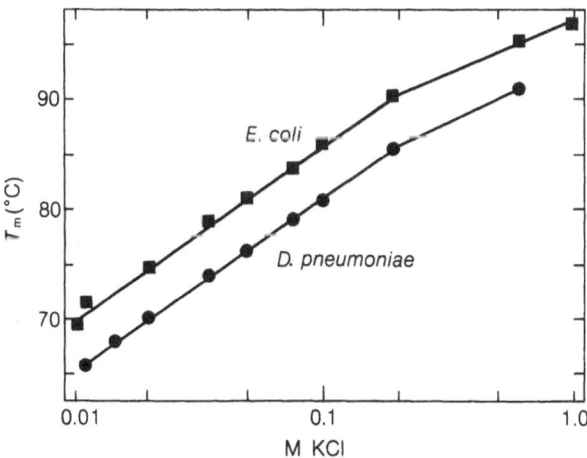

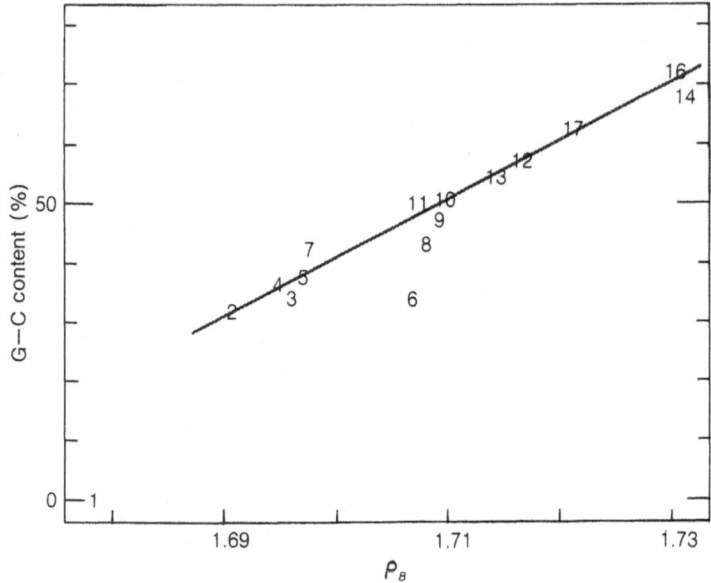

FIGURE 5.5. Buoyant density dependence (in CsCl) of DNA's on G-C content. Numbering of DNA's as in Fig. 5.3. (Data from Ref. 2.)

The most elegant way of measuring the length of a DNA molecule is by electron microscopy. It permits visualization and direct measurement of very long DNA molecules (chromosomes) at a level at which the validity of hydrodynamic methods is dubious. The error here is less than 10%.

5.1.4 HYBRIDIZATION

Hybridization is a technique used to detect homologous sequences between DNA and DNA or, more often, between DNA and RNA. For example, a radioactive-labeled RNA is incubated with denatured DNA (at about 20° below the T_m of the DNA; usually at 66° in 0.9 M NaCl + 0.09 M Na acetate for 24 hours). The RNA not hybridized is removed by ribonuclease degradation. The solution is then filtered through millipore filters (which retain only hybridized molecules) and counted. This permits one to estimate the amount of RNA bound to DNA. This is a bimolecular, second-order equilibrium reaction with very large association constants ($K = 10^{10}$). The heterogeneity of DNA sites, and of RNA's, however, requires a very careful study of the reaction in each case. It is very difficult to saturate all DNA sites, even at high RNA/DNA ratios. The stability of a hybrid depends on its G-C content, the ionic strength, and the length of the RNA (generally RNA's with chain length less than about

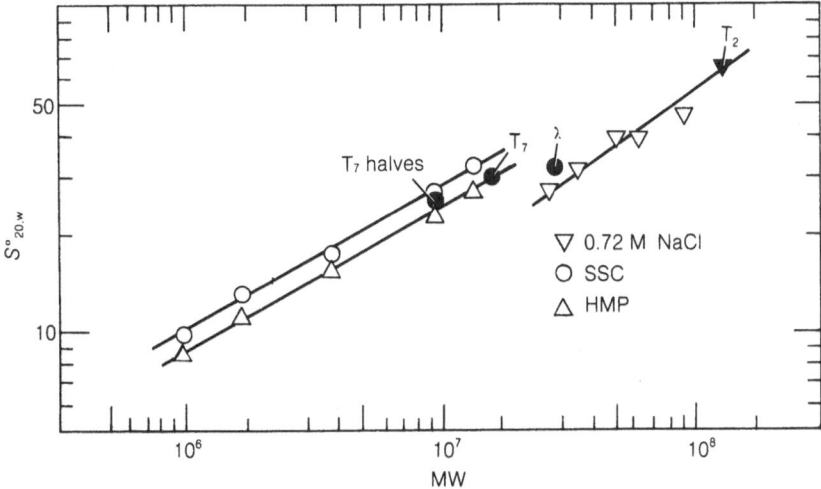

FIGURE 5.6. Relation between sedimentation constant $s^{\circ}_{20,w}$ and molecular weight of DNA of different size. (After Refs. 5 and 6.) HMP, 0.01 M phosphate buffer; SSC, standard saline citrate (0.15 M NaCl + 0.015 M Na-citrate).

20 do not hybridize under most experimentally feasible conditions). (See also Section 5.2.5.)

5.1.5 WHICH FORCES STABILIZE DNA STRUCTURE?

As shown in the previous sections, the stability of DNA is modified by a number of chemical and physical agents, such as variations in ionic strength and heat. Taking these results into account, one comes to the conclusion that essentially three forces stabilize the structure of DNA:

1. Hydrogen bonds between the complementary bases are apparently important for the specificity of base pairing, but these bonds cannot be the main stabilizing force of DNA for three reasons: (a) The energy of a hydrogen bond is of the order of 2 to 4 kcal/bond, which is not sufficient to account for the observed stability of DNA. (b) Hydrogen bonds break cooperatively under given conditions (e.g., at T_m). If a few hydrogen bonds in a DNA region open, the rest of them break virtually simultaneously. (c) Isolated bases (or nucleosides), even in very concentrated solutions, do not associate through hydrogen bonds, but "stack" up (see Chapter 6, Section 6.3), while their hydrogen-bonding capacity is preferentially satisfied by water.

2. From the above, apparently DNA stabilization is by stacking (hydrophobic) forces, which are due to the mutual interaction between the π-electrons of the aromatic bases (see also Chapter 6, Section 6.3). In a polynucleotide chain, this interaction yields a rather compact stack of bases in a

characteristic order. The restrictions of the sugar-phosphate backbone impose a quite narrow range of possible overlap angles between the bases (36° in DNA). The stack of bases forms a hydrophobic core, which favors hydrogen bonding between complementary strands. There is no free water within the double helix, a fact that favors associations with the complementary base rather than with water molecules. It is also very probable that the mechanism of nucleic acid synthesis is favored by the complementary bonding in a hydrophobic environment. Thus, while unwinding under certain energetic conditions is spontaneous, rewinding is not, and kinetic and entropic considerations enter the picture (unwinding, base specificity, etc.). Therefore, most complementary homopolynucleotides form double helices rather rapidly, whereas DNA renaturation is a very long and complex process, even under optimal conditions (see Section 5.2.5).

3. Ionic bonds between the negative phosphate groups and (positive) cations are important in that they reduce the electrostatic repulsion between the negative charges of the sugar phosphate backbone. If no counterions are present, double-stranded DNA becomes denatured. These cations may be metal ions, polyamines, or histones. Whereas metallic ions like Na, Mg, K, and Mn are abundant in all types of cells, polyamines are present in bacteria preferentially, and histones are found in higher organisms. Because polyamine and histone levels vary during the cell cycle, and because their specificity of binding is not absolute, it has been suggested that they act as general regulators of cellular mechanisms. We will return to this problem in Chapter 9.

5.2
Facts that May Require Amendments of Details of the Watson–Crick Theory

5.2.1 CLUSTERS AND "ISOSTICHS"

The nearest neighbor frequencies determined by Kornberg's group indicated that bases are not randomly distributed in DNA. This non-randomness, which is the foundation of the genetic message of DNA, is highly specific and is finally translated into protein sequences. Since three nucleotides code for one amino acid and only rarely, in a given protein sequence, is an amino acid flanked by the same amino acid, it was very surprising to find in many DNA molecules long runs (up to 15 and 20) of pyrimidines. Sometimes the same pyrimidine is found eight and ten times in a row (8). Only four amino acids are coded exclusively by pyrimidines (see Table 8.1). Pyrimidine "isostichs" (i.e., runs of the same length) are present in much larger quantities than would be expected from random distribution. Szybalski et al. (9) found long stretches of clusters of dCyt in T_7-DNA by selective hybridization with poly(U,G). These authors suggested that dCyt clusters might be the initiation signals for RNA polymerase. It should be pointed out that the isostichs and clusters are present as one sequence (or multiple sequences) per DNA molecule and are apparently characteristic for a given DNA.

5.2.2 ARE CLUSTERS DIFFERENT FROM THE REST OF DNA?

Numerous studies, mainly by optical methods, but also by X-ray fiber methods on synthetic polydeoxyribonucleotides, have shown that sequence analogues, e.g., poly(dA)-poly(dT) and poly(dA-T), have different properties, such as CD spectra and melting behavior (Figs 6.7 and 6.8). Similar differences have been observed with other all-purine–all-pyrimidine polymers compared to the alternating analogues (see Chapter 6, Section 6.1.3). The question arises whether such differences as the above in various segments of DNA are functionally significant. The answer is not known. Studies on the role of clusters in native DNA are limited by their low concentration (rarely more than 1 to 2%). Their presence has given rise to speculation about their biological role as punctuations or spacers in the genome.

5.2.3 A-T RICH DNA

Recently, two groups in Paris (10,11) have observed that DNA that is very rich in A-T ($>60\%$) does not show the B→A transition observed with G-C rich DNA. It is significant that this observation was made using two completely different methods, linear dichroism and X-ray diffraction and scattering. Since it is believed that DNA must change to the A form in order to be transcribed (12) (RNA–DNA hybrids are in the A form), these authors concluded that DNA regions containing primarily A-T pairs are not transcribed, but they may be the spacers between cistrons.

5.2.4 PRE-MELTING AND TRITIUM EXCHANGE. DOES DNA BREATHE?

If DNA is heated, its UV absorption changes abruptly at the T_m (Fig. 5.2). This transition is very narrow and cooperative (dependent on the length and homogeneity of the DNA) and, at 5° below and above the T_m, there is no further change in UV absorption. In 1962, Paleček (13), using a particularly sensitive method (polarography), discovered that even at room temperature some bases were already opening and their number increased gradually with temperature, up to the T_m, when the DNA opened completely. This observation, first received with reservation, was confirmed by the use of various other techniques. This phenomenon, called pre-melting, cast the first doubts on the model of DNA as a static molecule.

Simultaneously, Printz and von Hippel (14) showed by tritium exchange that all the hydrogen bond protons are slowly exchangeable (in a 100-second range). This important finding showed clearly that DNA has a dynamic structure with regions that opened and closed along the entire molecule—a sort of breathing. Also, by the use of formaldehyde (15) which traps free bases (Chapter 3, Section 3.2.2), it had been shown that the area of these regions, which open and close, is of the order of ten base pairs.

5.2.5 REPEATING SEQUENCES

Tandem repeating sequences. Under certain conditions of ionic strength and temperature, heat-denatured DNA will reassociate to reform double helical, biologically active DNA. This had been demonstrated by Marmur et al. (6,16, 17) on the transformation of *Hemophilus* DNA (these bacterial genomes have molecular weights of about one billion daltons). On the other hand, vertebrate DNA's cannot be reassociated easily. One of the reasons is that they are frequently heterogeneous populations of more or less degraded molecules. Furthermore, the molecular weights of vertebrate genomes are larger than bacterial genomes by a factor of about 1000.

Britten's group (18,19) observed that many vertebrate DNA's, especially if sheared to smaller pieces, will reassociate considerably faster than one would expect. This paradox, discovered in 1966, gave rise to the hypothesis that certain relatively short sequences of bases were repeated hundreds of thousands of times. It was also shown that the extent of reassociation of DNA was a measure of the evolutionary state of the species. Furthermore, heteroassociation (i.e., between DNA strands derived from different species) was shown to be a measure of the evolutionary relation between species. Finally, most eucaryotic DNA's contained one (or several) satellite components, which formed up to 10% of total DNA and showed very fast reassociation kinetics.

The techniques used for measuring reassociation kinetics are as follows: the dissociated DNA strands (generally obtained by heating to 100° for a few minutes, but also by bringing the DNA to pH 12) are incubated in a medium of a salt concentration greater than 0.01 M NaCl at a temperature about 25° below the T_m. The DNA concentration and incubation time must be chosen carefully if reassociation is to be efficient. The reaction is influenced by the size of the DNA; smaller (but not too small; see Section 5.1.4) DNA fragments reassociate faster than larger ones. Thus, sheared DNA preparations are frequently used. The degree of reassociation may be measured by the return of the UV absorption during reassociation. Another method is the selective binding of double-stranded DNA to hydroxy-apatite columns as a function of time.

Since the reassociation of two DNA strands will follow

bimolecular kinetics, the degree of reassociation will be determined by the time of reaction and the initial DNA concentration, c_0. The controlling parameter may be expressed as the product of the two, i.e., c_0t. Frequently, the c_0t value for half-association is used to characterize a given DNA sample. Reassociation kinetics are generally presented in a semilog graph (Fig. 5.7), fraction reassociated versus c_0t, which, ideally, should give a symmetric, sigmoid curve. The slope of the central part should be about 100, and any larger ratios indicate a heterogeneous reassociation.

The half-period of reassociation is proportional to the number of unit fragments present, i.e., the genome size. This has been verified experimentally, and, frequently, labeled *E. coli* DNA is used as an internal standard (genome size, 4.5 million base pairs). Evidently, this relation between c_0t at half-reaction and genome size implies non-repetitivity.

Although up to 30% of the DNA of higher organisms consists of repeating sequences (with a frequency ranging from a thousand to a million times), it is far from clear what purpose these regions serve. It is, however, certain that some genes are repeated in higher organisms, such as ribosomal RNA's or mitochondrial genes (21).

Palindromes in eucaryotic DNA's. Palindromes are sentences that read the same forwards and backwards. Thomas (22) used this term for the analogous nucleotide sequences in

FIGURE 5.7. Reassociation of DNA's from various sources. The DNA's sheared to about 400 base pairs, with the exception of phage and bacterial DNA's. (From Ref. 19; the data for rat liver are from Ref. 20.)

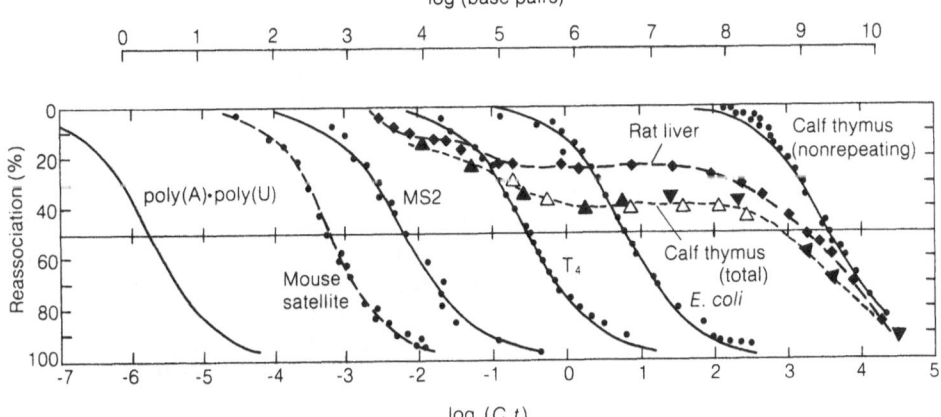

eucaryotic DNA's that are inverted tandem repeats, e.g.,

—N —A —B —C —D —E —F —E'—D'—C'—B'—A'—N'—
 | | | | | | | | | | | | |
—N'—A'—B'—C'—D'—E'—F'—E —D —C —B —A —N —

This sequence can, in principle, form a cruciform structure

which, if heat or alkali-denatured, will preferentially self-associate upon renaturation and form hairpins instead of the original double helix. Such hairpins (350–1200 base pairs long) have been observed by Wilson and Thomas (22) by electron microscopy in many eucaryotic DNA's, but they occur much less frequently than do tandem repeats.

Gierer (23) had proposed palindromic structures as recognition sites for DNA–protein interactions and, specifically for operon–repressor interactions. As will be seen below, certain nucleic acid sequences in such sites are, effectively, palindromes.

Restriction enzymes. Arber et al. (24) had postulated a model for the host-controlled specificity in *E.coli* (Fig. 5.8), involving specific sites (nucleotide sequences) on the viral DNA infecting a cell, which are recognized by a specific restriction endonuclease and are cleaved. Since such sequences may also occur in the host DNA, a methylase specific for these same sequences will place methyl groups on them and render them resistant against the restriction nuclease. This model has been found to be correct for *Hemophilus* (24) and, to a lesser extent, *E.coli*. Two enzymes have been isolated, an endonuclease and a methylase that recognize the same site of about four to eight base pairs. If methylase reaction has already occurred, placing two methyl groups on each strand, the endonuclease can no longer degrade the strand. This system of self-protection of the host explains how the cell can have a restriction endonuclease and not destroy its own DNA.

There are many different restriction enzymes in most cells (there are seven known in *E.coli*), which differ according to the base pair sequence they recognize (Table 5.1).

One of the most useful applications of restriction enzymes is their capacity to break large DNA's at specific points into fragments of defined size. Successive treatment of DNA

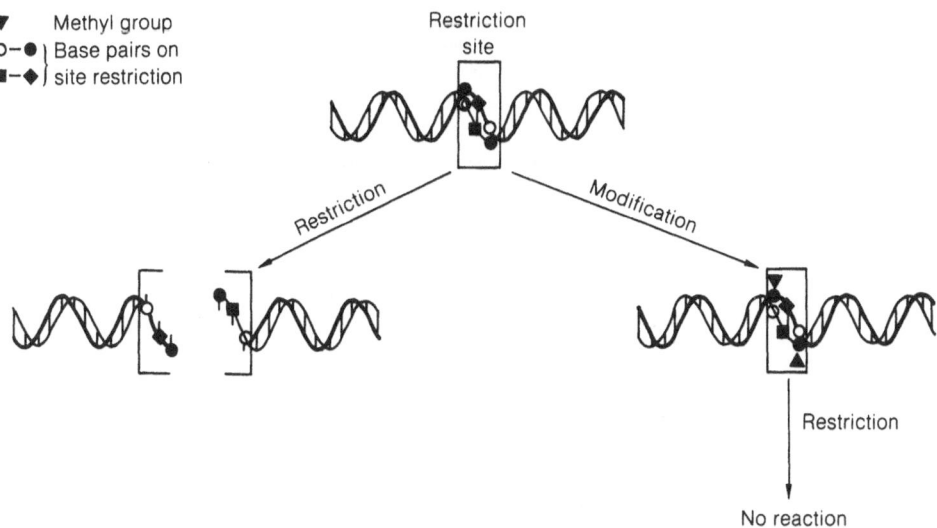

FIGURE 5.8. Scheme of the action of restriction and modification enzymes.

with different restriction enzymes produces cleavage maps and fragments suitable for sequence determination, localization of genetic loci, and isolation of operators (25). The potential of these enzymes is clear, and a large number of publications on their use have appeared recently.

The fact that the methylases place two methyl groups on two residues (always the same base in a given sequence, Ade or Cyt) led to the suggestion that the nucleotide sequence of the binding site might have dyadal symmetry and that the endonuclease that breaks both strands probably has dyadal symmetry. All sequences found so far confirm this supposition (Table 5.1), although the endonuclease attack does not always occur next to the modifiable base. In several cases (Eco RI and Eco RII), the restriction nuclease consists of two identical subunits, and the methylase is a monomer, which interacts with the dimer, attacking the same sequences.

Several of the restriction nucleases (e.g., Hind III, Eco RI, and Eco RII) do not cleave DNA at the same position on both strands but yield staggered breaks. These single-stranded regions can act as "sticky ends" (see p. 66). Evidently, another DNA fragment of different origin, with a single-stranded end with the same sequence (the sequences are symmetric), produced by the same restriction enzyme, can be hydrogen bonded at the site of the break. The two ends can then be sealed together with ligase, and a new DNA, consisting of two unrelated genomes, will have been

Table 5.1 Sequences Recognized by Restriction (R) and Modification (M) Enzymes[a]

Enzyme[b]	Sequence recognized[c]	R or M
Eco RI	<pre> * | —N—C—T—T—A—A—G—N—5′ 5′—N—G—A—A—T—T—C—N— | *</pre>	R,M

Eco RI′	<pre> |
	  —Y—Y—T—A—R—R—5′
	5′—R—R—A—T—Y—Y—
	              |</pre>
		R

Eco RII	<pre> * | —N—G—G—A—C—C—N—5′ 5′—N—C—C—T—G—G—N— | *</pre>	R,M

Eco (PI)	<pre> *
	  —T—C—T—A—G—A—5′
	5′—A—G—A—T—C—T—
	            *</pre>
		M

Hind II	<pre> * |
	  —C—A—R—Y—T—G—5′
	5′—G—T—Y—R—A—C—
	           | *</pre>
		R,M

Hind III	<pre> | *
	  —T—T—C—G—A—A—5′
	5′—A—A—G—C—T—T—
	   * |</pre>
		R,M

Hpa I	<pre> | —N—C—A—A—T—T—G—N—5′ 5′—N—G—T—T—A—A—C—N— |</pre>	R
Hpa II,Hap	<pre> | —N—G—G—C—C—N—5′ 5′—N—C—C—G—G—N— |</pre>	R
Hae III,Bsu x5	<pre> | —N—C—C—G—G—N—5′ 5′—N—G—G—C—C—N— |</pre>	R

a Data from Ref. 25.

b Enzyme abbreviations: Source and strain where enzyme was isolated, e.g., Hindii: enzyme II from Hemophilus influenzae strain D; EcoRII: Enzyme II from *E. Coli* strain R; Bsu: *Bac. subtilis*; Hpa: *Hemophilus parainfluenzae*; Hae: *Haemophilus aegypticus*.

c | Cleavage point; * methylated base; R: purine; Y: pyrimidine.

formed. This kind of experiment has been performed with bacterial and phage DNA's to introduce various new genetic markers. Even eucaryotic DNA pieces have been incorporated into these procaryotic DNA's and replicated *in vivo*. It is quite clear that these techniques open new dimensions in biochemistry and genetics, dimensions that may be extremely useful, but also quite dangerous. Thus, the transfer of DNA segments containing drug-resistance markers or viral enzymes requires great care to avoid spreading these potentially very dangerous DNA's. A recent appeal of involved scientists (26) clearly expressed concern, and warned against the potential biohazards of recombinant DNA molecules. In this connection, it should be pointed out that phage DNA's apparently involve less risk to humans, since they can only be propagated in the bacterial host. The phage structure limits the size of the DNA that can be packed into the phage head or capsid.

It is also of interest to determine whether these sequences, which are effectively palindromes of short chain length, can assume the cruciform conformation and thus be more readily recognized by the restriction enzyme. These problems are under study in several laboratories.

Operator–repressor interactions. Control and modulation of genetic expression by a reversible interaction of a specific repressor with a special site on the DNA molecule, called the operator, was proposed by Jacob and Monod (27). The first experimental work on such systems was concerned mainly with the *lac* and *lambda* repressors.

Lac repressor was isolated by Gilbert and Müller-Hill (28) and its properties extensively studied (29). It is a tetramer of four times 35,000 MW. It has not been possible to dissociate it reversibly into subunits. There are about ten repressor molecules in each *E.coli* cell. It is synthesized by the I gene (Fig. 5.9) and acts on the operator gene O. It can react with inducers (galactose analogues) directly or can be "derepressed" and thus released from the O gene. The operator region does not overlap the polymerase promotor P since RNA polymerase fixation is not blocked In the presence of *lac* repressor. It binds very tightly to one, and only one, site of the *E.coli* chromosome, the *lac* operator. *In vivo* and *in vitro* binding studies indicate a binding constant of about 10^{11} to 10^{13} (28,29). Binding to nonoperator DNA is about one million-fold smaller, and binding of the repressor to other DNA's decreases with increasing G-C content, suggesting that the *lac* operator must be rich in A-T sequences. The results of a genetic analysis seemed to indicate that the operator was symmetrical (30). The recently determined sequence of the *lac* operator (31) shows that it

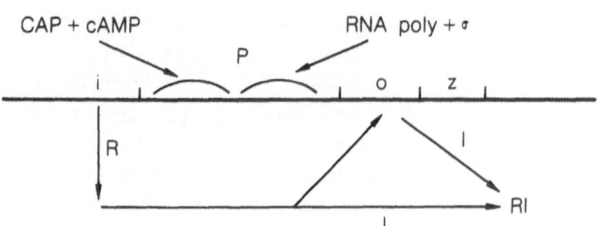

FIGURE 5.9. The lac-region in *E. coli*. RNA-polymerase region (P); operator (o); inducer (I); repressor (R).

is, indeed, symmetrical, and one can imagine formation of two loops.

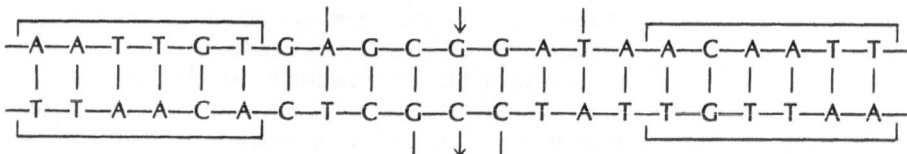

Operator–repressor association kinetics have provided valuable information about the mechanism of the control process. The dissociation constant of the operator–repressor complex is about 10^{-3} sec^{-1}. The dissociation constant is decreased by anti-inducers and increased by inducers as IPTG (isopropyl-1-thio-β-D-galactopyranoside). This permitted the identification of the "real" inducer, which is not lactose, but allolactose. Also, the ratio of forward and backward rate constants is about 10^{13}, which agrees with the binding constants measured *in vivo* and *in vitro*. Only the forward rate constant ($K = 10^{10}$ M^{-1} sec^{-1}) is excessively high. This cannot be accounted for by a normal diffusion-controlled process, the rate constant of which can be, at best, 1000 times lower.

Circular DNA—Cohesive Ends. Many viral DNA's, such as SV40 or papilloma virus, are circular. The *E.coli* chromosome is circular. DNA from phage lambda is linear, but both of its ends on opposite strands have single-stranded regions with an exactly complementary sequence of 12 nucleotides. These "sticky" or cohesive ends can hybridize to form a circular structure. In the T-phages, it has been shown that the double-stranded, linear DNA helix contains the same terminal sequence. With exonuclease III, the 3'-strand can be digested selectively, leaving single-strand cohesive ends that can anneal and form circles. The replicative form of the normally single-stranded DNA phage ϕX-174 is a double-stranded circle (see Chapter 4, Section 4.3).

It thus appears that many, if not all DNA's of viral or bacterial origin, even if linear in the vegetative form, circularize during replication and that the repetitive end sequences serve as locks to fuse the circles. Most of the closed circular DNA's have an additional feature: they are supercoiled, i.e., they lack a certain constant number of helix turns and will therefore compensate for this strain by supercoiling. They are thus packed more tightly. The supercoil structure of circular DNA's can be relaxed spontaneously by single-stranded breaks or gradually (Fig. 5.10) by intercalation of such aromatic dyes as ethidium bromide (32) (see Chapter 7).

5.3 What Do We Require from Biologically Operative DNA?

In the model of Watson and Crick, DNA is a uniform polyanion, 20 A wide and of great length, with its stacked bases turned inward. Only some of the functional groups of the bases stick out into the grooves. In the large groove: C^5 of the pyrimidines (methyl in Thy) and the N^7 and C^8 of the purines and an amino hydrogen of Ade and Cyt. In the small groove: N^3 of the purines, the C^2 keto group of Thy and Cyt and one hydrogen of the amine group of Gua (see Fig. 4.2). There is no apparent recognition site, except possibly a characteristic arrangement of bases and their specific functional groups in the grooves, to be read by an interacting protein. Steric, kinetic, and thermodynamic considerations do not support such recognition mechanisms. Certain

FIGURE 5.10. Reversal of supercoils in a closed, circular, supercoiled DNA helix. The Watson–Crick helix is presented as a single continuous line (a). The use of intercalating dyes (perpendicular bars) will decrease the number of supercoil turns (b); at the equivalence point, where the accumulated untwisting due to the intercalating dyes will equal the number of initially present (a) superturns, the DNA molecule will form a circle (c). Further intercalation will cause further untwisting, thus forming supercoil turns in the opposite, left-handed sense (d), (e). The use of DNase, which introduces one single-strand break, relaxes the original (a) DNA molecule directly to the circle (c), with one break in one of the two strands. (From Ref. 32.)

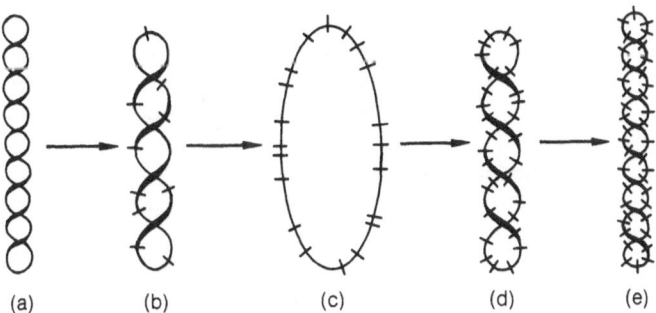

(a)　　(b)　　(c)　　(d)　　(e)

sites, which are probably only present infrequently, may be distinguished by other criteria, such as subtle differences in secondary structure (8–11). Repetitive sequences, which would offer the possibility of cooperativity and a different secondary structure, or palindromes, which could assume the cruciform structure, would be good candidates for such recognition sites.

In addition, the different functions of DNA (replication, transcription) and the different steps in these processes (initiation, termination) as well as their control (repressors, effectors, etc.) require that each of these sites be very clearly different and unique. A modulated structure (14,15) that "breathes," and allows the formation of specifically distinguishable sites (even if they are metastable), would apparently favor rapidity of a recognition process.

5.4 What Do We Still Not Know?

All the problems of DNA structure are clearly not yet solved. Although there is considerable evidence that DNA is not a uniform helical structure, we are far from knowing the details. Geneticists and biochemists have clearly established that, in many instances, only very small regions of DNA are involved in the expression of certain functions. These regions must necessarily be different from their neighboring regions (and from the rest of the DNA) in order to be recognized. The problem frequently becomes one of technique: physiocochemical methods, although highly specific, are always macroscopic. In the study of minute, microscopic differences in DNA structure, one must cope with a large amount of nearly uniform DNA (>99%), with all that this implies in the accumulation of errors. For this reason, model compounds are frequently used (Chapter 6), but extrapolation of results from models to local sites in native DNA, which are perhaps not longer than one helix turn (ten base pairs), is dubious, although necessary in research. Spectacular advances in the enzymology of nucleic acids and DNA still have not provided a full physical description of the physiological mechanisms involving DNA. For instance, there is very limited information concerning the way polymerases act upon DNA, in which groove they act, how they open the DNA molecule, etc.

At present, DNA research is increasingly directed toward the study of nucleic acid–protein interactions (see Chapter 9) (33), as a means of probing the nature of the intermediates in the whole process of genetic expression.

References

1. Marmur, J. and Doty, P. (1962). *J. Mol. Biol.*, **5**, 109–18.
2. Schildkraut, C., Marmur, J., and Doty, P. (1962). *J. Mol. Biol.*, **4**, 430–43.
3. Samejima, T. and Yang, J. T. (1965). *J. Biol. Chem.*, **240**, 2094–2100.

4. Gratzer, W. B., Hill, L. R., and Owen, R. J. (1970). *Eur. J. Biochem.*, **15**, 209–14.

5. Rubinstein, I., Thomas, C. A., and Hershey, A. D. (1961). *Proc. Nat'l Acad. Sci., U.S.A.*, **47**, 1113–22.

6. Doty, P., Marmur, J., Eigner, J., and Schildkraut, C. (1960). *Proc. Nat'l. Acad. Sci., U.S.A.*, **46**, 461–76.

7. Klotz, L. C. and Zimm, B. H. (1972). *J. Mol. Biol.*, **72**, 779–800.

8. Chargaff, E. (1968). *Prog. Nucleic Acid Res. Mol. Biol.*, **8**, 297–334.

9. Szybalski, W., Kubinski, H., and Sheldrick, P. (1966). *Cold Spring Harb. Symp. Quant. Biol.*, **31**, 123–28.

10. Bram, S. (1971). *Nature, N. B.*, **232**, 174–76; Bram, S., and Tougard, P. (1972) *Nature, N. B.*, **239**, 128–31.

11. Pilet, J. and Brahms, J. (1972). *Nature, N. B.*, **236**, 99–100.

12. Arnott, S., Fuller, W., Hodgson, A., and Prutton, I. (1968). *Nature*, **220**, 561–65.

13. Paleček, E. (1969). *Prog. Nucleic Acid Res. Mol. Biol.*, **9**, 31–74.

14. Printz, M. P. and Von Hippel, P. (1968). *Biochemistry*, **7**, 3194–3206.

15. Von Hippel, P. and Kwok-Ying, W. (1971). *J. Mol. Biol.*, **61**, 587–613.

16. Marmur, J. and Lane, D. (1960). *Proc. Nat'l. Acad. Sci., U.S.A.*, **46**, 456–61.

17. Marmur, J., Rownd, R., and Schildkraut, C. L. (1963). *Prog Nucleic Acid Res. Mol. Biol.*, **1**, 231–300.

18. Waring, M. and Britten, R. J. (1966). *Science*, **154**, 791–95.

19. Britten, R. J. and Kohne, D. E. (1968). *Science*, **161**, 529–40.

20. Szala, S. and Chorazy, M. (1972). *Acta Biochim. Pol.*, **19**, 235–50

21. Kohne, D. E. (1970). *Quart. Rev. Biophys.*, **3**, 327–75.

22. Wilson, D. A. and Thomas, C. A. (1974). *J. Mol. Biol.*, **74** 115–44.

23. Gierer, A. (1966). *Nature*, **212**, 1480–81.

24. Arber, W. (1974). *Prog. Nucleic Acid Res. Mol. Biol.*, **14**, 1–37

25. Murray, K. and Old, R. W. (1974). *Prog. Nucleic Acid Res. Mol Biol.*, **14**, 117–85.

26. Berg, P. et al. (1974). *Proc. Nat'l. Acad. Sci., U.S.A.*, **71**, 2593–94

27. Jacob, F. and Monod, J. (1961). *J. Mol Biol.*, **3**, 318–56.

28. Gilbert, W. and Müller-Hill, B. et al. (1967). *Proc. Nat'l. Acad Sci., U.S.A.*, **56**, 1891–98; (1967). *ibid.*, **58**, 2415–21, (1968) *ibid.*, **59**, 1259–64.

29. Riggs, A. D. and Bourgeois, et al. (1968). *J. Mol. Biol.*, **34**, 361-64; 365–68; (1970). *ibid.*, **48**, 67–83; **51**, 303–14; **53**, 401–17.

30. Smith, T. R. and Sadler, J. R. (1971). *J. Mol. Biol.*, **59**, 273–305 **62**, 139–69.

31. Gilbert, W. and Maxam, A. (1973). *Proc. Nat'l. Acad. Sci. U.S.A.*, **70**, 3581–84.

32. Crawford, L. V. and Waring, M. (1967). *J. Mol. Biol.*, **25**, 23–30

33. Von Hippel, P. and McGhee, J. D. (1972). *Ann. Rev. Biochem.* **41**, 231–300.

6 Model Systems for Nucleic Acids

6.1
Polynucleotides

Most of the polynucleotide work has been designed to determine physical parameters of simplified systems of nucleic acids. Studies have been performed to determine the contributions of base pairing, base stacking, and chain length to structure, as well as the effects of sequence.

Like double-stranded DNA, synthetic polynucleotide complexes containing complementary bases show cooperative melting curves, i.e., the property studied (UV absorption, optical activity, viscosity, etc.) shows a sharp transition at a specific temperature. This T_m depends on ionic strength and, to a certain degree, on pH. If the temperature is increased above the T_m, however, a further continuous variation in the property studied will be observed. It is only in recent years that an understanding of such noncooperative phenomena is emerging. Here the studies of oligomers by CD and NMR were of critical importance.

Several additional results have arisen from these studies. Polynucleotides can not only form Watson–Crick helical double-stranded complexes but may also form helical structures between themselves which can have more than two strands, as well as non-Watson–Crick base pairs, like the complex poly(I)·poly(A)·poly(I). Furthermore, numerous polymers of base and sugar analogues have been prepared and studied.

Although the structure of polynucleotides is far from being completely understood, the interest in these model systems has somewhat declined. The reason is rather simple: the obvious experiments have been performed, and the others are much more complicated. Much of the earlier work on polynucleotides has been reviewed by two groups (1,2) in 1967.

6.1.1 HOMOPOLYNUCLEOTIDES

Almost all polyribonucleotides have been synthesized by polynucleotide phosphorylase (3). This unusual enzyme polymerizes nucleoside diphosphates to 3′-5′ linked poly-ribonucleotides, liberating one mole of inorganic phosphate. The reaction is reversible and favors the phosphorylation reaction. The detailed mechanism is not yet clear, but it appears that, after an initial phase of synthesis of long polymers, an equilibrium between synthesis and degradation establishes itself and polymers of reasonably uniform length are formed. The enzyme polymerizes riboside diphosphates and a number of analogues, such as 2′-O-methyl, 2′-chloro and 2′-fluoro-riboside diphosphates and even, but to a much lesser extent, arabinoside diphosphates, but not 2′-deoxyribonucleoside diphosphates. The preferential 3′-endo conformation seems to be mandatory, since 2′-endo deoxy compounds are not substrates.

On the other hand, virtually all base analogues are polymerized, even the nonaromatic dihydrouridine diphosphate, the exception being nucleotides in which the base is locked into the syn conformation, such as 8-bromopurine analogues or formycin. Many other polynucleotides have been prepared by chemical modification of preformed polymers.

Since polynucleotide phosphorylase does not use a template (in some cases a primer, which is incorporated into the nascent polynucleotide chain, is used), essentially homopolymers have been synthesized and studied. Heteropolymers have random distribution of the incorporated bases.

To obtain ribopolymers of defined sequence, transcription of deoxyribopolymers by RNA polymerase has to be employed.

Polydeoxyribonucleotides are more difficult to obtain than the corresponding ribopolymers. There are only a few detailed studies on these polymers. Deoxyribopolymers are synthesized by DNA polymerase, either de novo, yielding homopolymers or alternating heteropolymers—poly(dA-dT) and poly (dI-dC). Alternatively, they are synthesized by using chemically synthesized short templates of defined sequence, which are copied by a "slipping mechanism" by DNA polymerase into long polymers of defined sequence.

Poly(U) and poly(dT). Polyuridylic acid [poly(U)] shows little secondary structure at room temperature apart from base stacking. Below 15°, a stable secondary structure is formed. This form is very probably (4) a hairpin structure, i.e., the poly(U) chain bends back upon itself. This is a difficult topological problem, since such a form has obligatorily antiparallel strands, which are not equivalent, i.e., there is no dyad axis in the plane of the base pair, nor parallel to the helix axis. A double-stranded structure with parallel strands and a dyad axis parallel to the helix axis would be much more satisfying.

Poly(C) and poly(dC). Poly(C) forms a double-stranded structure upon protonation, with a shared proton between the two N^3 of the bases (Fig. 6.1). In contrast to the double helices of DNA and heteropolynucleotide complexes, the strands of this helix have the same polarity. The dyad axis is therefore not perpendicular to the helix axis (as in DNA), but parallel to it; in fact, it coincides with the helix axis. Much evidence (5) shows that the acid form of poly(C) is ionic strength dependent, and its thermal dissociation can take place either by uptake or loss of a proton, depending on the solvent conditions. This yields a rather strange, bell-shaped phase diagram (Fig. 6.2). At neutral pH, poly(C) forms a single-stranded, stacked helical structure, which, even at 100°, is not completely unstacked.

Poly(dC) is partly stacked at neutral pH and forms an acid double helical structure analogous to that of poly(C), except that, in the case of poly(dC), the transitions take place about two pH units higher. No really satisfactory explanation yet exists for this phenomenon.

Poly(A) and poly(dA). Similar to poly(C), poly(A) forms a double-stranded parallel protonated helix, which, however,

FIGURE 6.1. Planar arrangements in polynucleotide self-structures (A, C, and G).

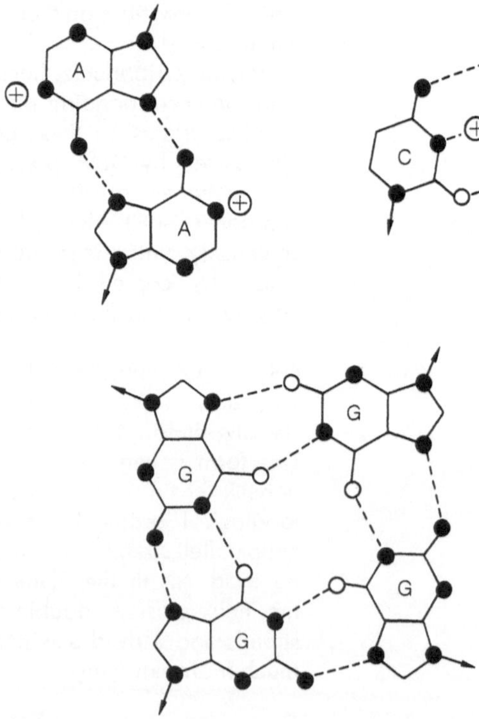

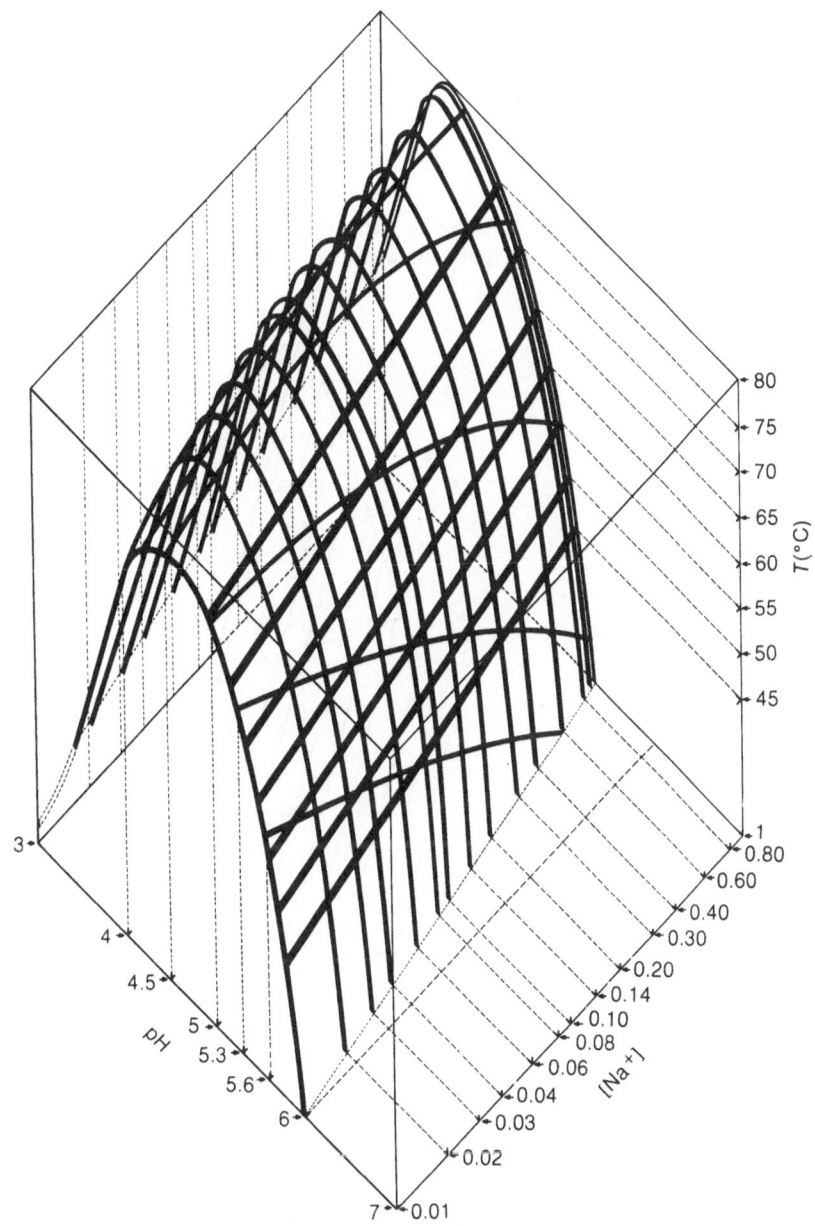

FIGURE 6.2. Phase diagram of poly(C) as a function of pH, temperature, and ionic strength. (From Ref. 5.)

does not use the extra protons for its hydrogen bonds. The X-ray structure (6) of acid poly(A) (Fig. 6.1) was determined long ago, but, recently, new X-ray data (7) suggested another double helical, possibly unprotonated form. The demonstration of this unprotonated, double-stranded poly(A) in solution has not been conclusive (8), but a thermodynamic analysis has permitted a detailed description of the poly(A) phase diagram (Fig. 6.3). Poly(dA) also forms an acid double-helical structure, which has not been studied extensively.

FIGURE 6.3. Phase diagram of poly(A) as a function of pH, temperature, and ionic strength. (From Ref. 8.)

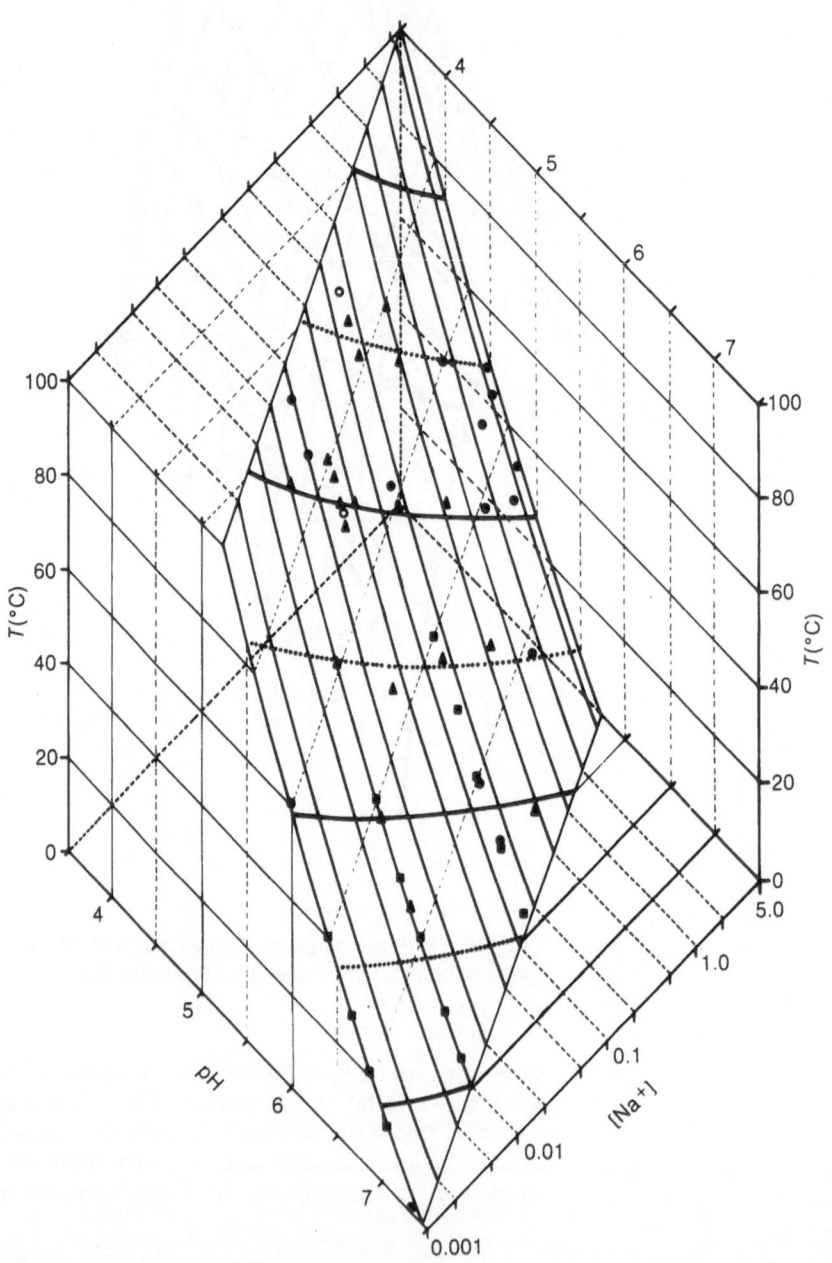

Poly(I) and poly(G) and their deoxyanalogues. The structure of poly(I), which is poly(G) less the N^2 amino group, has long been assumed to be triple stranded (9), but, from recent solution (10) and X-ray (11) data, it has been quite conclusively shown that poly(I) forms a four-stranded, parallel helix. This structure (Fig. 6.1) is also found in poly(G) (11,12), in analogy with the gels formed by Guo and its analogues (13) (see below). Methylation (14) of poly(G) or poly(I) on either N^1 or N^7 breaks up their secondary structure; this implies that these sites participate in the hydrogen bonding of the structure.

Poly(dG) and poly (dI) have similar, but less stable structures.

The four-stranded poly(I) structure surprisingly shows a 2'-endo puckering in contrast with all other ribopolynucleotides studied so far (Table 6.1) and also in contrast with the 3'-endo puckering of the three-stranded polynucleotide complexes.

6.1.2 POLYNUCLEOTIDE COMPLEXES

Soon after the discovery of polynucleotide phosphorylase (3) and the synthesis of the first polymers, it was observed that two complementary polymers, such as poly(A) and poly(U) or poly(I) and poly(C), will interact to form a complex (1,2,15). The technique employed is the classical Job plot or continuous variation experiment (Fig. 6.4). A series of mixtures of constant total concentration, but different mole ratios, of the two substances to be investigated (in this case polynucleotides) are studied by measuring a parameter that changes with the interaction; the most widely used parameter is absorption at a characteristic wavelength

FIGURE 6.4. Continuous variation experiment (Job plot) or mixing curve between two complementary polynucleotides, here poly(A) and poly(U). Note that the shapes of the curves are quite different at different wavelengths. (After Ref. 16.)

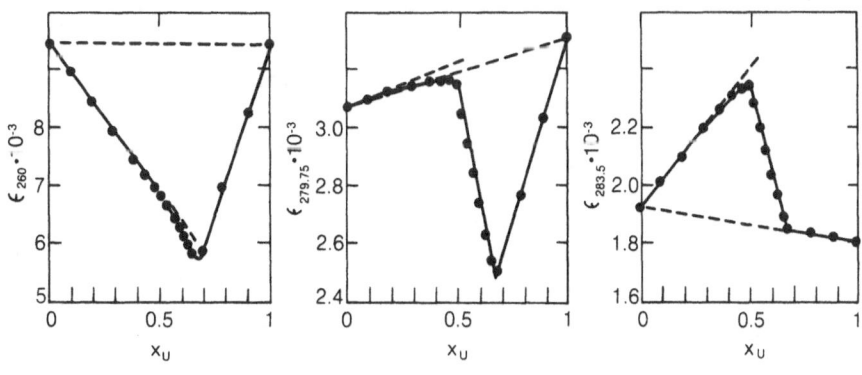

(Fig. 6.4). If no interaction occurs, a straight line between the absorption of A and U will be observed. Interaction between the two compounds gives rise to hypochromicity (see Chapter 4), and the intersection of the lines of observed absorbances will indicate the stoichiometry of the complex formed. Under suitable conditions, more than one complex may be observed.

Poly(A)·poly(U). Poly(A) and poly(U) can interact to form a double-stranded structure with Watson-Crick hydrogen bonding, and also a triple-stranded form, with the second poly(U) strand bound in a Hoogsteen pair (15).

Massoulié (15) has studied the interaction between poly(A) and poly(U) in detail and has established a detailed phase diagram at neutral and acid pH (Fig. 6.5). At neutral pH, there are regions where both the double- and the triple-stranded complex can coexist (region I). At lower ionic strength (region II) only poly(A)·poly(U) exists, whereas at high ionic strength and high temperature (region III) only the triple-stranded complex is stable. At temperatures above the limits of these three zones (region IV),

FIGURE 6.5. Phase diagrams of the interaction between poly(A) and poly(U). The different structures at pH 7 are a function of temperature and ionic strength (From Ref. 15.)
   I: Coexistence of poly(A)·poly(U) and poly(A)·poly(U)·poly(U)
   II: Only poly(A)·poly(U)
   III: Only poly(A)·poly(U)·poly(U)
   IV: Dissociated strands poly(A) and poly(U)

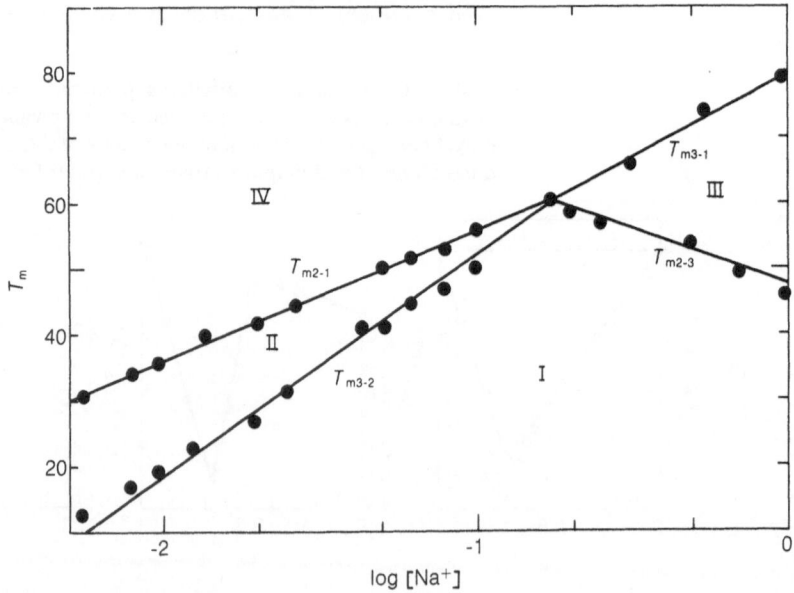

only the dissociated strands of poly(A) and poly(U) will be present. At acid pH there is competition with the acid form of poly(A). This destabilizes the complex and gives rise to hysteresis loops (17), due to the higher stability of acid poly(A).

The argument over the orientation of the strands in poly(A)·poly(U)·poly(U) and the way the second Urd is paired with Ado has probably been settled by the work of Leng (18), who showed that the poly(U) strand bends back upon the poly(A) and is, therefore, parallel to the poly(A) strand. This favors pairing of the second Urd with Ado in the Hoogsteen manner. Recently, new X-ray data (19) have confirmed this geometry. It should be noted that the best X-ray pictures ever obtained (20) are those of poly(A)·poly(U) which is in the A form [Fig. 6.6(a)].

Recently (22), complexes between substituted poly(A) and poly(U) have been made that permit only the Hoogsteen pair to be formed. The experimental proof for the existence of the Hoogsteen pair in polymeric form at neutral pH is of obvious importance.

Other A- and U (or T)-containing complexes. We shall consider here ribo- and deoxyribopolymers with alternating sequences, such as poly(dA-dT), as well as homopolymer interactions, such as poly(dA)·poly(dT) or poly(rA)·poly(dT). This series of compounds had been designed and studied

Table 6.1 Polynucleotide Structures Studied by X-Ray Fiber Diffraction. Helical Parameters, Sugar Conformations, and Possibilities of A↔B Transition[a]

Polynucleotide	Helix Symmetry	Genus	Sugar puckering	A↔B Transition	h[b] (A)	γ[c] (deg.)
A-DNA	11/1	A	3'-endo	Yes	2.56	20.2
B-DNA	10/1	B	3'-exo	Yes	3.37	6.3
C-DNA	28/3	B	2'-endo	Yes	3.32	−7.8
D-DNA(poly(dAT) and poly(dGC))	8/1	B	3'-exo	Poor	3.03	−16.7
poly(dA)·poly(dT)	10/1	B	3'-exo	Never	3.24	−8.0
poly(dA)·poly(dT)·poly(dT)	12/1	A	3'-endo	Never	2.26	7.2–9.1
poly(dG)·poly(dC)	11/1	A	3'-endo	Poor		
A-RNA (poly(A)·poly(U))	11/1	A	3'-endo	Never	2.82	17.4
A'RNA (poly(I)·poly(C))	12/1	A	3'-endo	Never	2.00	12.5
poly(A)·poly(U)·poly(U)	12/1	A	3'-endo	Never	2.04	10.2–12.0
poly(A)·poly(U)·poly(U)	11/1	A	3'-endo	Never	2.04	11.7–13.6
poly(I)·poly(I)·poly(I)·poly(I)	23/2	B	2'-endo	Never	3.41	8.9

[a] After Arnott et al. (19–21, 27).

[b] h: Helix rise per base pair (A).

[c] γ: Tilt of base pair to normal of helix axis (deg.).

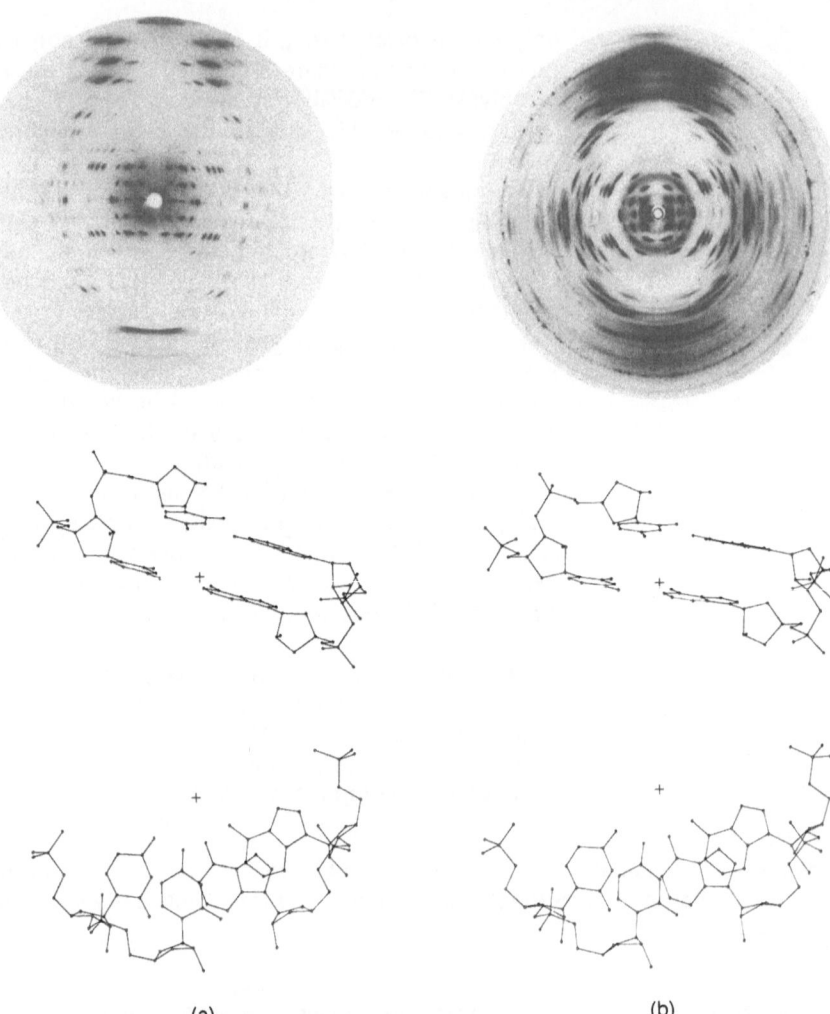

(a)   (b)

FIGURE 6.6. X-ray fiber diffraction patterns of A-RNA (left)
[poly(A)·poly(U)] and A'-RNA (right) [poly(I)·poly(C)]. Fibers are
tilted 75° to the X-ray beam. Relative humidity is 92% (left) and
75% (right). (From Refs. 19 and 20.) *Bottom:* projections
perpendicular and along the helix axis of the two forms. Note
the large tilt in both cases as well as the displacement of the
helix axis before the base pairs. (Compare Fig. 6.12.)

to obtain information on sequence effects and on the differences between RNA and DNA. Although these are simplified models, significant conclusions can be drawn from
their study.

   All these polymers and polymer complexes possess organ-
ized helical structures at room temperature and at the usual
salt concentrations. Their structures are generally double

helical. Crystallographic studies on poly(dA-dT) have shown that it occurs in a B-like form (Table 6.1), with a poor B→A transition. Hybrid ribo-deoxyribo polymer complexes occur in the A form, which indicates that the 3'-*endo* conformation of the riboside imposes its conformation.

The problem of sequence dependence of different properties in DNA was first indicated (23) by the very large difference in the optical properties of poly(dA)·poly(dT) and poly(dA-dT) (Fig. 6.7) despite their identical base composition. This difference is similarly accentuated in other DNA-sequence analogues. (see Section 6.1.2). On the other hand, poly(dA-dT), with a regular alternating ATAT sequence, is an ideal model compound for many studies. The CD spectra of both compounds, however, show large variations at pre-melting temperatures, which indicate structural changes.

The alternating polymers, like poly(dA-dT), possess a repeating self-complementary sequence that can fold back on itself. Branched helices (hairpins) can thus form and have been observed by electron microscopy. These polymers therefore show some distinctive features. The branched structure will rearrange at intermediate temperatures to

FIGURE 6.7. CD spectra of (a) poly(dA-dT) at 5°, 35°, and 55°; (b) poly(dA)·poly(dT) at the same temperatures. (After Ref. 24.)

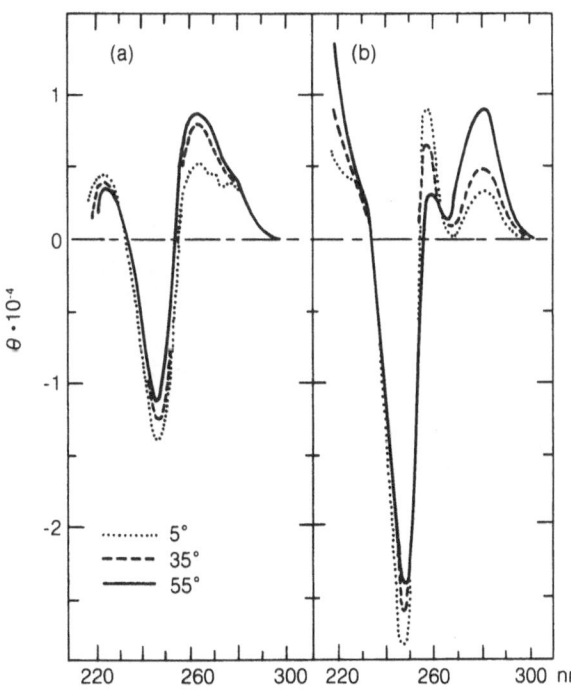

form a continuous helix. Kinetic studies (25) on these alternating structures have provided important clues on the mechanism of helix formation.

Studies of the ionic strength dependent properties of the various polymeric complexes (Fig. 6.8) have established several important principles. All the double helical structures show $\Delta T_m = 20° \pm 2°$ for a tenfold increase in ionic strength (Fig. 6.8), exactly as do all double-stranded DNA's and RNA's. Three-stranded polymers generally show $\Delta T_m \sim 30°$. Records (26) has discussed in great detail the thermodynamic implications of this observation.

There are also a few problems: as can be seen from the data summarized in Fig. 6.8, poly(A)·poly(U) shows T_m's about 10° lower than the alternating polymer poly(A-U). The converse is true for the pair poly(dA-dT) and poly(dA)·poly(dT). Although both ribopolymer complexes are probably in the A form (or one of its subclasses), the deoxypolymers possibly have two different helical structures, although both are variations of the B form (27).

The T_m values at a given ionic strength are thus in the order:

FIGURE 6.8. Ionic strength dependence of A- and U (or T)-containing polynucleotide complexes.

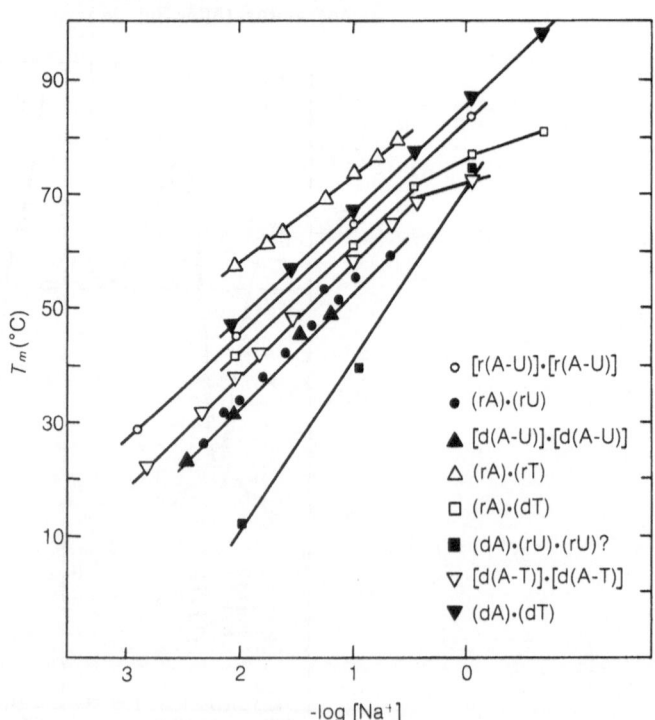

$$(rA)\cdot(rT) > (dA)\cdot(dT) > (rA\text{-}rU) > (rA)\cdot(dT) > (dA\text{-}dT) >$$
$$(rA)\cdot(rU) = (dA\text{-}dU) > (dA)\cdot(rU)$$

Note that the hybrid homopolymer complex always melts at a lower temperature than either the all-ribo or the all-deoxyribo complex; also, the all-ribo complex is thermally the most stable. It should be noted that DNA-RNA hybrid structures melt at a temperature about 4° lower than that for DNA itself, which, in turn, has a considerably lower T_m than a double-stranded RNA with the same base composition. Similar observations are made for the G-C and I-C series (see below).

Poly(G)·poly(C) and poly(I)·poly(C) and their deoxyanalogues.
Both poly(G) and poly(I) form 1:1 complexes with poly(C) under appropriate solvent conditions (neutral pH, high ionic strength) (28). Although the complex poly(G)·poly(C) does not denature even at high temperatures, poly(I)·poly(C) shows a strong ionic strength dependence on T_m (Fig. 6.9), analogous to that observed for other two-stranded complexes $(\Delta T_m/\Delta \log[Na^+] = 20°)$ (26). Both complexes rearrange upon acid titration to form triple stranded complexes

FIGURE 6.9. Ionic strength dependence of C- and G (or I)-containing polynucleotide complexes.

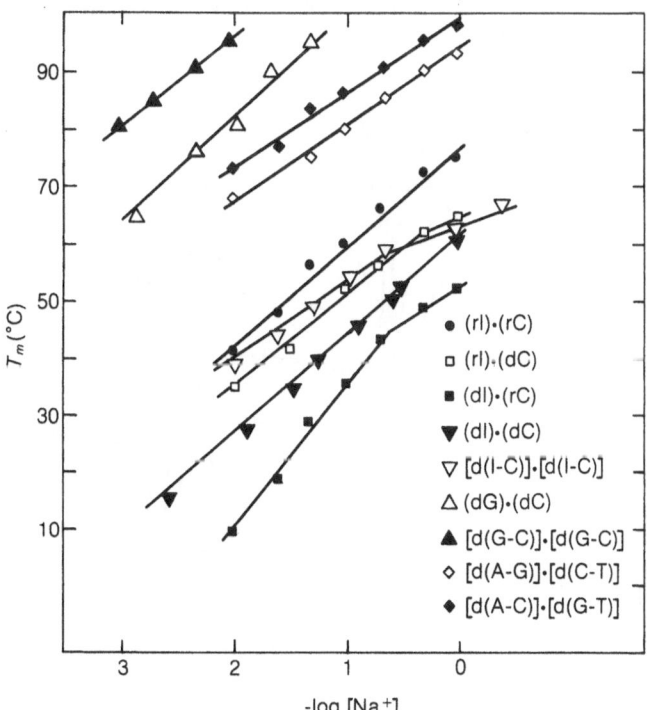

T_m(°C)

-log [Na⁺]

- (rI)·(rC)
□ (rI)·(dC)
■ (dI)·(rC)
▼ (dI)·(dC)
▽ [d(I-C)]·[d(I-C)]
△ (dG)·(dC)
▲ [d(G-C)]·[d(G-C)]
◇ [d(A-G)]·[d(C-T)]
◆ [d(A-C)]·[d(G-T)]

containing two poly(C) strands, the second poly(C) being connected by the additional proton (28). These transitions are, again, ionic strength dependent (Fig. 6.10). At lower pH, poly(I)·poly(C)·poly(C⁺) forms a double-stranded poly(I). poly(C⁺), which is probably Hoogsteen paired (Fig. 6.11), no analogous transition is observed for poly(G)·poly(C)·poly(C⁺), the pH being so low that the polymers precipitate.

Poly(I)·poly(C) forms Watson–Crick pairs in the A form (11-fold helix) or, in the presence of excess salt, a 12-fold helix (form A′) [Fig. 6.6(b)] (20).

In the I–C series, all the ribo- and deoxyribopolymers have been synthesized and all their combinations studied. Again the all-ribo-double helix is that with the highest T_m (Fig. 6.9). The order of T_m's of these polymers is, however, not quite as expected:

$$(rI)·(rC) > (rI)·(dC) > (dI)·(dC) > (dI)·(rC),$$

FIGURE 6.10. Phase diagram of poly(I) and poly(C) complexes. (a) (rI)·(rC); (b) (rI)·(rC)·(rC⁺); (c) (rI)·(rC⁺). (From Ref. 28.)

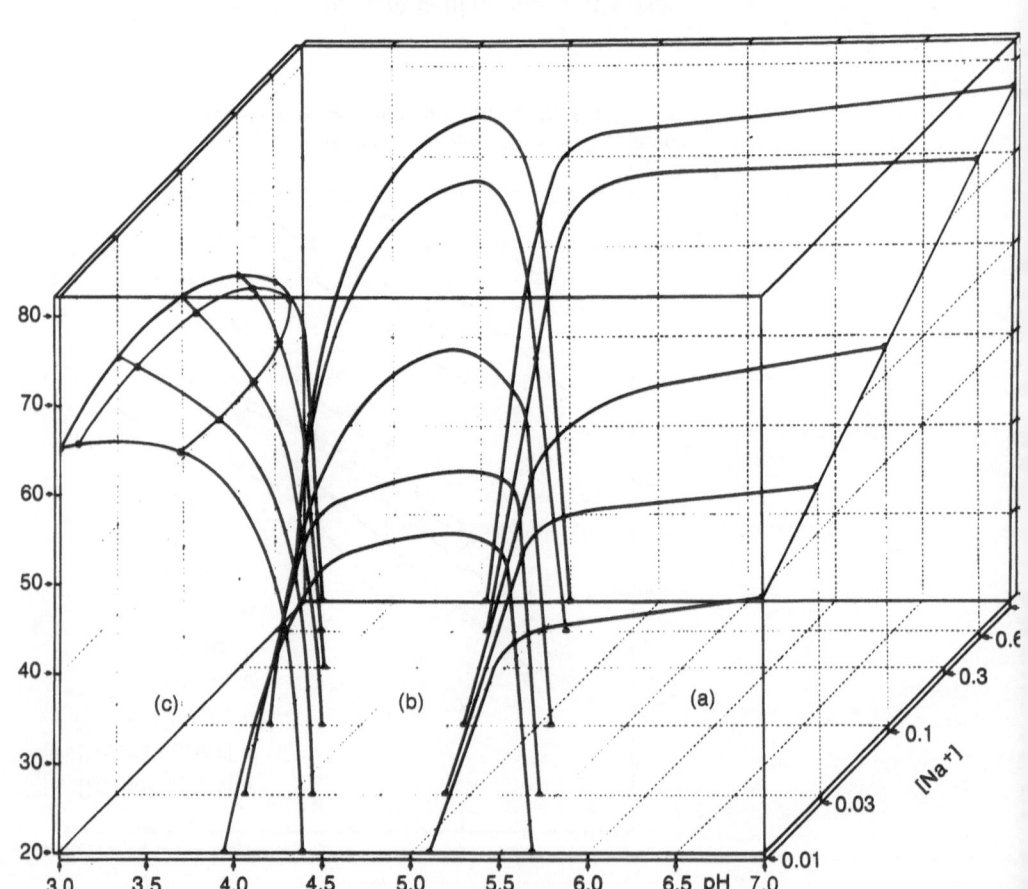

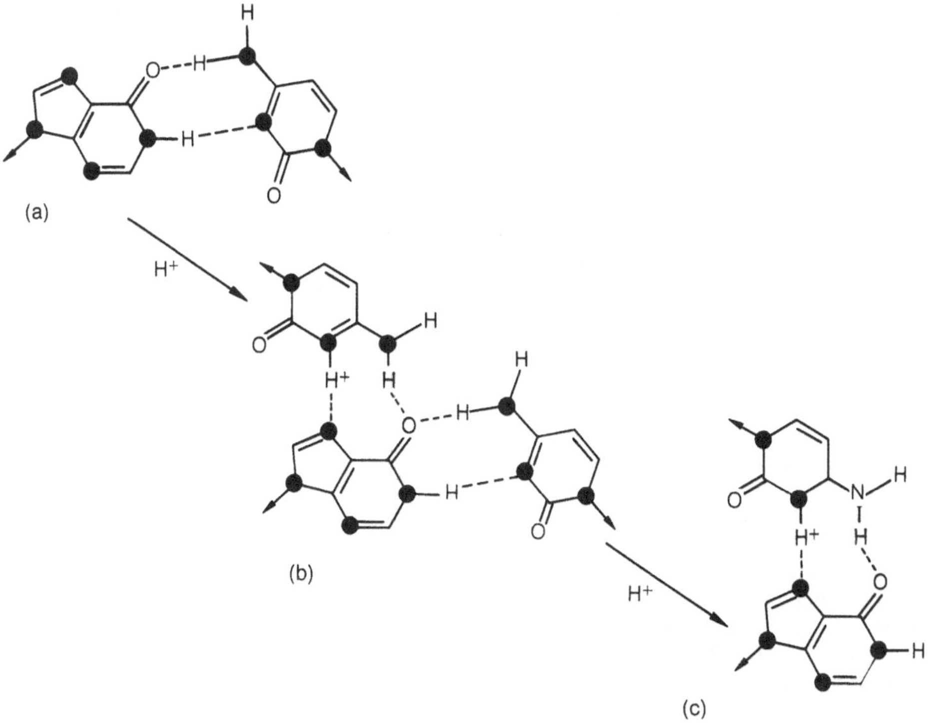

FIGURE 6.11. Disproportionation of poly(I)·poly(C) during acid titration. (From Ref. 28.) (See Fig. 6.10.)

the hybrid complex (rI)·(dC) being thermally more stable than the deoxypolymer complex.

The alternating deoxypolymers poly(dI-dC) and poly(dG-dC) are thermally more stable than the homopolymer complexes, an apparently general finding.

Arnott's group is currently studying all these polynucleotide complexes by X-ray fiber diffraction, and the results will clarify some ambiguities. Thus, poly(dG)·poly(dC) in fibers was found to be preferentially in the A form. The alternating polymers poly(dA-dT) and poly(dG-dC) show a rather unusual D form, with eight base pairs per turn (Table 6.1). Although the D→A transition is observed at low relative humidities, the A form is metastable, and the transition is never complete, returning rapidly to the D form (27).

It should be noted that, in the D-helix, the base pairs are displaced backwards relative to the helix axis (Fig. 6.12), i.e., in the opposite direction to that observed for the A form. It should be emphasized that it is the sugar puckering that defines the helical genus.

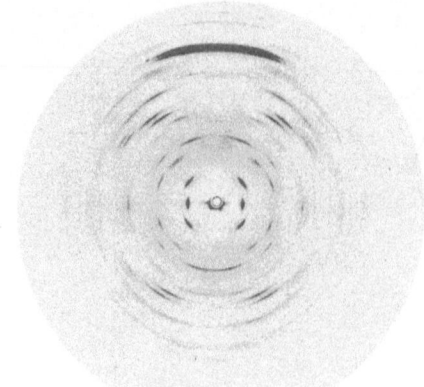

(a)

(b)

(c)

FIGURE 6.12.
X-ray fiber diffraction pattern
(75% rel. humidity, fibre tilted
72° to x-ray beam) of the
D form of DNA assumed by
poly(dA-dT) and poly(dG-dC);
(a) and projection of two
nucleotides (TpA). (b) View
perpendicular to the helix axis;
note the negative tilt. (c) View
along the helix axis. Note that
the bases are located behind the
helix axis in this structure in
contrast to that observed in the
A and B forms of DNA and RNA
(see Fig. 6.6 and Fig. 4.7). (From
Ref. 27.)

On the other hand, all three-stranded helices, even those
of the deoxy family, are in the A form.

Further X-ray studies of such complexes will give some
valuable information on their fine structure, which will be
important for our understanding of sequence effects in DNA
(see Chapter 5, Section 5.2).

Some other polynucleotide complexes. Polyxanthylic acid [poly(X)] forms complexes of various stoichiometries (1:1, 1:2, 2:1) with poly(U) and/or poly(I) (30).

Poly(I) forms a triple-stranded complex with poly(A).

Other sequence pairs. Khorana's group (31–33) has developed a method of synthesizing DNA polymers of known sequence. Chemically synthesized oligonucleotides were used as templates for DNA polymerase. In this case, a "slipping mechanism" permitted the copy and repeat copy of these to form long polymers (chain length > 200). These polymers have proved to be extremely useful in studies of sequence effects on the physical properties of DNA, and they have been studied extensively, especially by Wells (32,33).

It was found that characteristic differences existed between polydeoxyribonucleotides that contain only purines in one strand and pyrimidines in the other strand, as compared with those containing the same bases, but in alternating sequences. Figures 6.8 and 6.9 contain T_m data (33) of such sequence pairs containing A + T, G + C, and A, G, C and T, either as homo-purine·homo-pyrimidine or alternating sequences. As a general rule, the mixed sequences show slightly higher thermal stability (the A-T pairs are exceptions). The polymers with the higher T_m always have a lower buoyant density in CsCl. The CD spectra show great variations between the pairs (Fig. 6.13). Also, the X-ray re-

FIGURE 6.13. CD spectra of poly(dG)·poly(dC), poly(dG-dC), poly(dA-dG)·poly(dC-dT) and poly(dA-dC)·poly(dG-dT) (Redrawn from Ref. 29.)

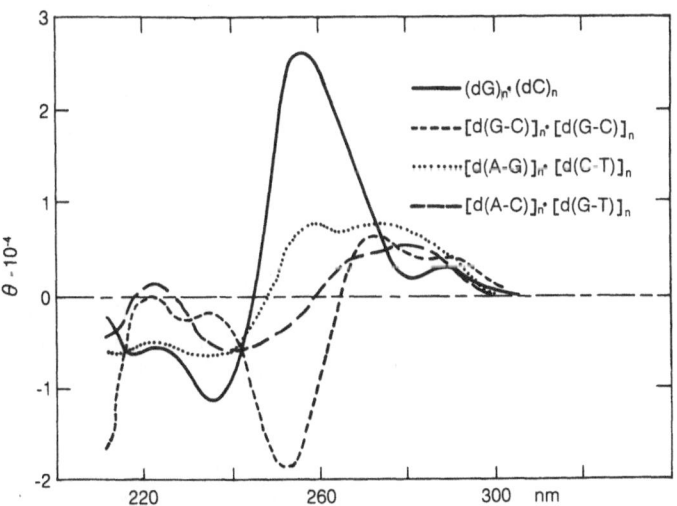

sults obtained for synthetic polydeoxyribonucleotides have
been shown to differ for homologous pairs (27,34).

6.1.3 COMPARISON BETWEEN RIBO- AND DEOXYRIBOPOLYNUCLEOTIDES

The structural differences between DNA and RNA are well
documented, but poorly understood. If differences between
double-stranded DNA and RNA are already considerable,
they are even larger in the single-stranded homopolynucleotides. Some of the most striking examples have been reviewed (1), but our understanding of these differences has
been rather slow.

Single-stranded polydeoxyribonucleotides are, as a rule,
less stacked and structured than their ribo counterparts.
They frequently show different titration behavior [e.g.,
poly(dC) forms the double helical acid form two pH units
higher than poly(C) under the same experimental condi-
tions] and optical properties.

Numerous explanations for the greater stability of ribopolymers have appeared. For example, a hydrogen bond
between the 2'-OH and the neighboring phosphate group
has been proposed (35). Studies on 2'-O-methylated polynucleotides (36,37), which are more highly structured than
the polydeoxyribonucleotides, refute this explanation, however. To elucidate this problem, other analogues with fluorine, chlorine, acetate, or azide at the 2'-position have been
made; all of them were found to be more stable than the
deoxypolymers. No conclusive explanation for this fundamental problem of nucleic acid chemistry has been found,
but there are indications that the polarity of the 2'-substitu-
ent may be a determinant for the pucker of the sugar,
which, in turn, determines helix geometry.

6.1.4 MECHANISM OF HELIX FORMATION

The availability of synthetic polynucleotides of defined sequence has prompted many workers to study the mechanism
of helix formation. Equilibrium (thermodynamic) and dynamic (kinetic) studies show that the formation of the first
base pair (nucleation) is the rate-limiting step (38,39) and
that the equilibrium constant for nucleation is about 1000
times lower than that for the formation of any following
base pair. This means that subsequent base pair formation is
greatly favored, once the first pair has been established and
that few intermediates will be formed. [We shall not discuss
in detail this rather complicated field, which depends heav-
ily on mathematical models and methods.]

6.1.5 CONCLUSIONS ON POLYNUCLEOTIDES

The studies on polynucleotides have brought forth much
useful information on the properties of nucleic acid bases

in a polymeric structure. Except for some unusual bases (e.g., xanthine), these studies present a large body of evidence in favor of the Watson–Crick base pairing scheme. Three-stranded structures, however, can be formed under certain circumstances by the addition of Hoogsteen pairing.

Crick had suggested, and the tRNA sequences confirm, the existence of "wobble" pairs (see Chapter 8). A few precise data on polymeric wobble structures have appeared so far (40). It may well be that single wobble pairs (like I-U) may be stabilized in the ribosome–tRNA–messenger complex, but, outside the complex, they would have stabilities too low to be observed under the usual experimental conditions. On the other hand, it has been shown (41) that noncomplementary bases will loop out preferentially and not remain in the polymer stack.

6.2 Oligonucleotides

Michelson (42) was the first to observe that oligo- and even dinucleotides showed hypochromicity without double helix formation. This first indication that base-base interactions not involving hydrogen bonds may play an important role in the stabilization of nucleic acids has now been amply confirmed. The phenomenon is called stacking and is due to the interaction of the· π electrons of the neighboring bases. These stacking interactions strongly influence optical parameters (decrease in absorption, increase in ORD and CD) (43).

Dinucleoside phosphates are considered the simplest cases for the study of stacking interactions. They have been studied principally by optical methods, mainly to obtain information concerning the effect of sequence on optical properties. The UV absorption or CD of MpN should differ from that of NpM; this should also permit their identification. Tinoco's exciton theory (44) has been tested on these model compounds, and reasonably satisfactory results have been obtained for some homodinucleotides, in particular ApA. Temperature studies on these model compounds were designed to define the stacking forces between bases. It would be expected that, at high temperature, the stacking contributions would essentially vanish. Finally, these studies have also permitted interaction enthalpies between bases to be estimated, using a two-state model of a temperature-dependent equilibrium between stacked (S) and unstacked (U) forms (Table 6.2, Fig. 6.14):

$$S \rightleftharpoons U$$
$$K = f_u/f_s = (1 - f_s)/f_s$$

and

$$-RT \ln K = \Delta H - T\,\Delta S$$

In principle, the unstacked CD spectrum should be that of the constituent nucleotides, and this is generally the

Table 6.2 Apparent Thermodynamic Parameters as Determined from Optical Melting Curves

Compound	ΔH (kcal/mole)	ΔS (cal/mole·deg)	ΔF (25°) (cal/mole)
ApA	−8.0	−28	−400
ApC	−6.1	−21	−600
ApU	−6.7	−24	−300
GpA	−6.1	−22	−200
GpU	−5.6	−28	− 50
CpA	−7.0	−24	−400
CpC	−7.5	−25	−700
CpU	−6.8	−24	−200
UpA	−7.0	−25	100
UpG	−6.7	−23	−300
UpC	−5.4	−19	−275
UpU	−6.1	−22	−500

case. Subtracting the monomer spectra from the observed CD spectrum of the dinucleotide should thus gives a difference spectrum, representing the temperature-dependent fraction of an interaction spectrum. This, in turn, could be computed, if the necessary parameters were known. Again, the agreement is satisfactory for a few homodinucleotides (43,44) [Fig. 6.15(a)], while, for hetero-pairs, large deviations are observed [Fig. 6.15(b)], indicating that the theoretical parameters are not adequate. The geometry used for these computations is that of DNA, i.e., parallel bases at a 36° angle.

For several reasons, PMR studies on dinucleotides are difficult. In the case of a homodinucleotide, the difference between the chemical shifts of the protons of the 3'-bound nucleoside and the 5'-bound one are rather small, particularly for the sugar protons, and thus difficult to identify. Analysis of the protons of the bases is easier, since one expects changes in chemical shifts of stacked bases (see also Section 6.3.1) due to the ring current. Still, many studies (46) have appeared in which it was concluded that all dinucleotides were in a right-handed *anti-anti*–conformation. This conclusion appears to be justified for the homooligonucleotides, but possibly may not extend beyond that. To avoid overlap of the peaks of the two nucleosides, one of them might be completely deuterated. This has been done (47), and the study on ApA confirmed the previous optical data. The synthesis of such a deuterated dinucleotide is, however, not a simple affair.

NMR analyses of the sugar residues have indicated that the conformation of the two riboses in dinucleoside phos-

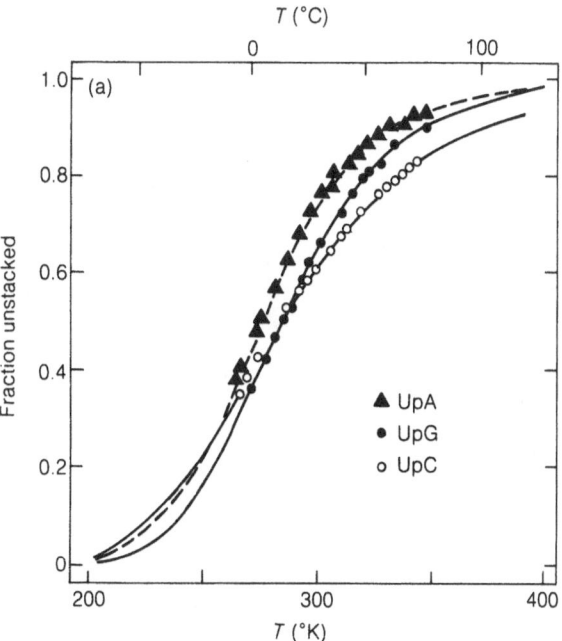

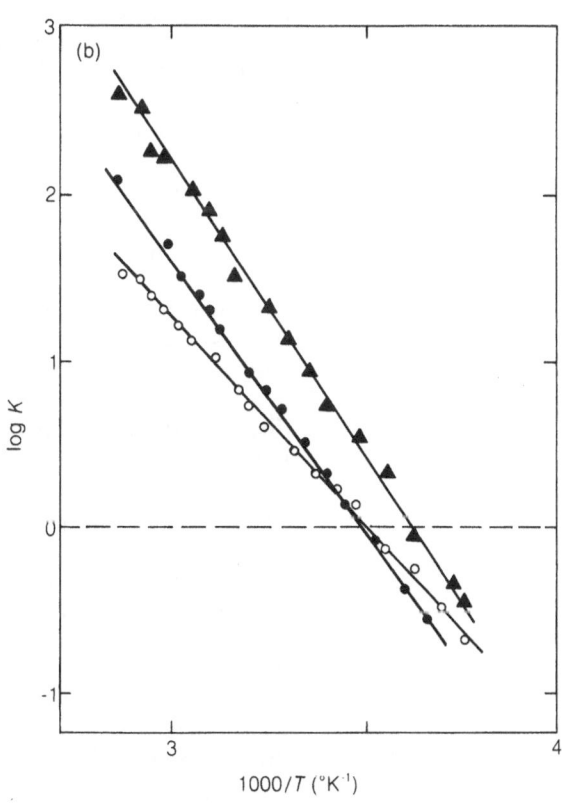

FIGURE 6.14.
Melting curves of UpA, UpC, and UpG, at pH7. (a) Computed curves and experimental points; (b) van t'Hoff plot of the data in (a).

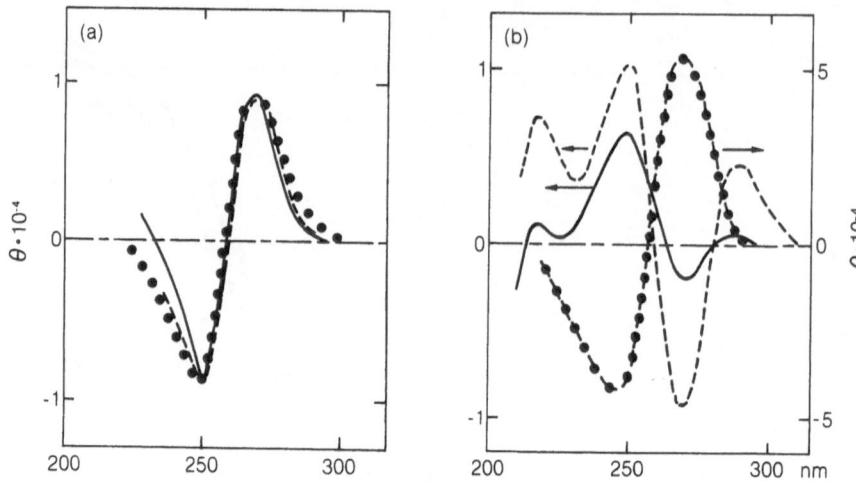

FIGURE 6.15. CD spectra of ApA (a) and GpU (b) (Refs. 43, 44, 45.) pH 7 in water (——), pH 1 in water (- - -). Theoretical curves are computed from exciton theory (• - • - •).

phates are different. Both sugars tend to approach the S conformation at higher temperatures.

There has been some disagreement between authors using CD and those using NMR concerning the validity of various conclusions. It should be kept in mind that CD measures perturbations in the chromophores (bases) and that variations in the sugar-phosphate backbone will probably not be seen in the CD spectrum. On the other hand, NMR data may give important information on the sugar-phosphate conformations, but much less precise data on base-base interactions. Thus, unstacking will cause a continuous variation in CD and UV absorption, but rather abrupt effects in the chemical shifts of the base protons.

A major, unsolved problem is the difference between ribo- and deoxyribodinucleotides, the latter being generally less stacked. Although it has been established that the nature of the 3'-bound sugar determines the structure and stability of a dinucleotide (48), the reason for this is not yet clear. There are many interesting studies on this problem, similar to those on polynucleotides, but no conclusive explanation has been offered. The most plausible finding to date is the difference in sugar puckering between ribo- and deoxyribonucleoside diphosphates, which seems to be determined by the polarity of the 2'-subsituent; these differences in puckering do not seem to be as pronounced as those seen in nucleic acids or polynucleotides (46).

Homooligonucleotides with chain lengths longer than two have been studied (43). In the stacked form, the bases of an oligonucleotide of chain length n are either in (n-1) contacts with neighboring bases or with the solvent (the two terminal bases). Optical properties will mainly be determined by the (n-1) interactions, and, therefore, the contribution of the end effects will decrease with chain length. The hypochromism and the CD per residue decrease approximately with $1/(n-1)$ (38).

The NMR data indicate that the sugar residues of the terminal nucleotides are different from and more flexible than those inside the stack. More refined NMR data will increasingly contribute to the understanding of the forces and conformations that stabilize oligonucleotides.

6.3 Association of Monomers in Solutions and Crystals

6.3.1 SELF-STACKING

In solutions normally used for absorption work (below ten millimolar), nucleosides generally follow the Lambert–Beer Law, i.e., the signal is linearly proportional to the concentration. Above these concentrations, deviations from linearity are observed (Fig. 6.16). From this hypochromism, it was deduced that base–base interactions (stacking) occur. Generally, the decrease in extinction coefficient is gradual over relatively wide concentrations, and no sharp breaks are seen (Fig. 6.16). It is thus difficult to extract precise information from such data. Two other techniques have here been largely employed: vapor pressure osmometry and NMR.

Vapor pressure osmometry is based essentially on Raoult's Law. In an isolated thermostated container, two drops are placed upon two thermistors, one drop contains the solvent, the other the solution with the compound to be studied. The two drops will have a different chemical po-

FIGURE 6.16. Changes in molar extinction coefficient with concentration for adenosine and guanosine. Ado in water: data from Ref. 49; Ado in CHCl$_3$: data from Ref. 52; Guo in 0.1 M KCl solution: data from Ref. 13.

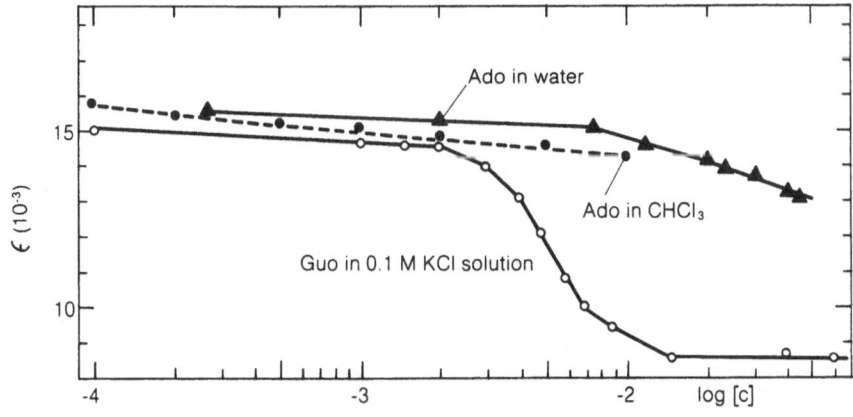

tential and, thus, vapor pressure. This difference is proportional to the number of molecules in the solution. An appropriate electric circuit compensates for the difference in tension, the signal of the compensating current being proportional to the amount of solute. If, at a given concentration, the number of molecules changes due to association or dissociation, the signal will also change. By this technique one can measure the osmotic coefficient $\phi = M_1/M_n$, where M_1 is the monomer molecular weight and M_n the number average molecular weight of the associate. If no association occurs, the osmotic coefficient will be 1. The technique permits the determination of relatively small molecular weights ($<$3000–4000).

The large decrease in the osmotic coefficient of virtually all bases, nucleosides, and nucleotides above about 0.05 M has been interpreted as reflecting an association of the bases in stacks, the size of which depends on the concentration (49,50). From the changes in chemical shifts of characteristic protons (H^8 and H^2 in purines, H^5 and H^6 in pyrimidines, and $H_{1'}$ in ribose) (51) a head-to-tail stack was deduced as the most probable form. It should be noted that, in all these studies, Guo behaves quite differently, as discussed in the following section.

6.3.2 GEL FORMATION

There is a particularly strong tendency for Guo and its derivatives to associate. It does not follow the Lambert–Beer Law (Fig. 6.16), but, in contrast to other bases that form monomer stacks, Guo first associates to form a planar tetramer (Fig. 6.1); two such tetramers then form an octamer and these stack up to form highly viscous, and even solid pseudopolymers, which are highly birefringent. This tetramer arrangement has been shown to characterize all the gels of Guo and its analogues (13). On the other hand, any change on the functional groups of the bases inhibits gel formation. No other bases or their derivatives have been found to form gels.

6.3.3 BASE PAIRING IN SOLUTION

In contrast with Guo derivatives, which self-associate in aqueous solutions to form hydrogen bonded pseudopolymers, all other nucleosides self-associate only to a very limited extent by stacking (see Section 6.3.1). To obtain information on the planar homologous or heterologous associations between bases, one must use non-polar solvents that do not associate (hydrogen bond) with the bases (dimethyl sulfoxide, chloroform). These solvents maximize hydrogen bonding between the bases and make self-stacking less probable. Since nucleosides are frequently poorly soluble in these solvents, these studies are usually done using free bases or their aliphatic derivatives (e⁹Ade, 1-cyclohexyl-Ura, m¹Cyt, e⁹Gua) (52,53).

Spectroscopic techniques are generally used to detect associations between bases (NMR, IR, UV). These studies have shown that Ade and Ura derivatives, on one hand, and Gua and Cyt derivatives, on the other, interact much more strongly with each other than with themselves or other bases to form hydrogen bonded heterodimers. Since in non-polar solvents the amino protons of the bases do not exchange, they can be studied by NMR techniques. Large downfield shifts of these protons or the displacement of the amino or N-H stretch frequencies have been observed. Similarly, the small UV hypochromism caused by the association between the bases (Fig. 6.17) has been interpreted as evidence for the 1:1 association between complementary bases in a planar fashion. In contrast, no other associations except those between A and U or G and C (or I and C) have been observed.

6.3.4 MIXED CRYSTALS OF BASES AND NUCLEOSIDES

The isolation of crystalline complexes containing monomer purine and pyrimidine derivatives and the determination of the base paired structure by X-ray crystallography has shed

FIGURE 6.17. (a) Association of 9-ethyladenine with 1-cyclohexyluracil in chloroform; absorption changes and oscillator strength are shown. (After Ref. 53.) (b) Association of 2′,3′-benzylidine-5′-trityl-guanosine with 2′,3′-benzylidine-5′-tritylcytidine in chloroform. 0.01, 0.0075, and 0.005 are the total concentrations present in the mixtures for the three series. (After Ref. 52.)

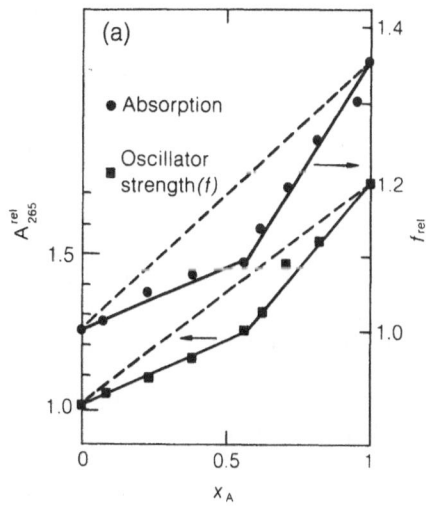
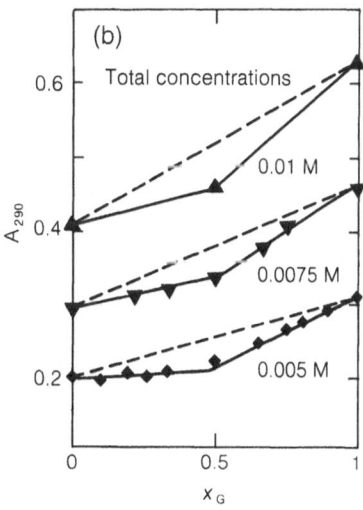

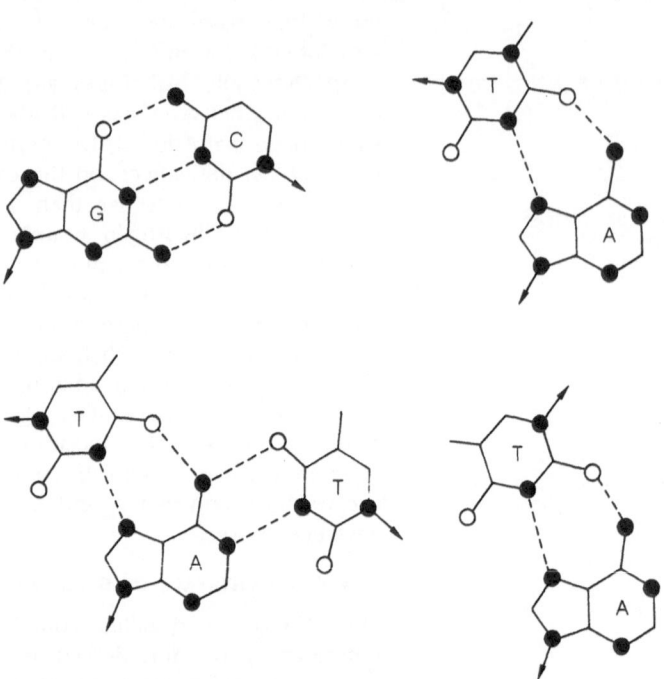

FIGURE 6.18. Base pairing arrangements in mixed crystal structures (From Ref. 56.)

light on the association possibilities of these compounds. Originally, these complexes were designed to verify the Watson–Crick base pairings to a higher resolution than is possible with fiber X-ray diffraction studies. The results of the first of these studies by Hoogsteen (54) were a surprise. The crystal complex between m^9Ade and m^1Thy does not form the Watson–Crick base pair. A hydrogen bond between the amino group of Ade and the 4-keto group of Thy is present, as in the Watson–Crick pair, but the second bond is formed between the N^3 of Thy and the N^7 of Ade (Fig. 6.18). The second surprise was the inverted Hoogsteen pair (55) with a hydrogen bond between the amino group of Ade and the 2-keto group of Ura or Thy. This Hoogsteen pair and its inversion have been found to be exclusively present in single crystal studies. Only if a 1:2 A:U crystalline complex is formed are the Watson–Crick and Hoogsteen pairs (and/or their reversed forms) observed (Fig. 6.18) (56).

A third surprise was the single crystal complex of G-C (57). Its Watson–Crick pairing is not surprising, since a G-C Hoogsteen pair would require an additional proton. The sugar conformation was, however, in the *syn* conformation

Over the main entrance of the Corpus Domini Cathedral in Orvieto (Italy), one finds examples of parallel and anti-parallel multi-stranded helices. The middle of the three columns is three stranded and has its triad (threefold) axis parallel to the helix axis. The four-stranded helix on the left and the six-stranded helix on the right have dyad axes perpendicular to the helix axis, giving rise to a large and a small groove. In both cases, one strand in the small groove and one in the large groove have no polarity. Note the flexibility of the helices (top).

in the dGuo and *anti* in the br⁵dCyd molecule. Only protonated G-C complexes would be expected to be in the Hoogsteen form, but they have not been studied.

In this connection, it is also interesting that in the crystal structures involving nucleosides the conformation around the glycosidic linkage is always *anti*, except in the only

Guo-containing complex. Here the orientation is *syn* (57), which agrees with the high flexibility of this bond in solution.

6.3.5 CONCLUSIONS

We have seen that monomers can associate in many ways and form a variety of complexes. In this complicated picture some rules appear:

1. In aqueous solutions, bases have a tendency to interact by base-base stacking interactions, the hydrophilic functional groups (amino, keto, and N-H) probably being solvated.
2. In nonaqueous, aprotic solutions, in which the substituents are not involved in hydrogen bonds with the solvent, hydrogen bonds between the bases are strongly favored, giving rise to dimers or complementary base pairs.
3. In crystals, even in the presence of small amounts of water, both inter- and intramolecular hydrogen bonds are observed and also hydrogen bonding to solvent molecules and stacking in characteristic patterns.
4. Guanosine is an exception, because it has a greater number of functional groups (Fig. 6.1) and hydrogen bond donor and acceptor sites than any other base. The tetramer has two hydrogen bonds per base, whereas, in other planar associations, at most three hydrogen bonds for two bases are found (the G-C pair); thus, several functional groups are free to form H bonds with the solvent. This tetramer is then capable of self-association by base–base interactions stabilized by additional, vertical hydrogen bonds.

References

1. Michelson, A. M., Massoulié, J., and Guschlbauer, W. (1967). *Prog. Nucleic Acid Res. Mol. Biol.,* **6**, 83–141.
2. Felsenfeld, G. L. and Miles, H. T. (1967). *Ann. Rev. Biochem.,* **35**, 407–48.
3. Grunberg-Manago, M. (1963). *Prog. Nucleic Acid Res. Mol. Biol.,* **1**, 93–133.
4. Thrierr, J. C., Dourlent, M. and Leng, M. (1971). *J. Mol. Biol.,* **58**, 815–30.
5. Guschlbauer, W. (1967). *Proc. Nat'l. Acad. Sci., U.S.A.,* **57**, 1441–48; (1975). *Nucleic Acids Res.,* **2**, 353–60.
6. Rich, A., Davies, D. R., Crick, F. H. C., and Watson, J. D. (1961). *J. Mol. Biol.,* **3**, 71–86.
7. Finch, J. F. and Klug, A. (1969). *J. Mol. Biol.,* **46**, 597–600.
8. Vetterl, V. and Guschlbauer, W. (1972). *Arch. Biochem. Biophys.,* **148**, 130–40.
9. Rich, A. (1958). *Biochim. Biophys. Acta,* **29**, 502–11.
10. Thiele, D. and Guschlbauer, W. (1973). *Biophysik,* **9**, 261–77.

11. Arnott, S., Chandrasekaran, and Martilla, C. M. (1974). *Biochem. J.*, **141**, 537–43; Zimmerman, S. B., Cohen, G. H. and Davies, D. R. (1975). *J. Mol. Biol.*, **92**, 181–92.

12. Guschlbauer, W. (1972). *Jerusalem Sypm.*, **4**, 297–310.

13. Chantot, J. F. and Guschlbauer, W. (1972). *Jerusalem Symp.*, **4**, 205–16; Tougard, P., Chantot, J. F., and Guschlbauer, W. (1973). *Biochim. Biophys. Acta*, **308**, 9–17.

14. Michelson, A. M. and Pochon, F. (1966). *Biochim. Biophys. Acta*, **144**, 469–80.

15. Massoulié, J. (1968). *Eur. J. Biochem.*, **8**, 428–38, 439–47.

16. Massoulié, J., Blake, R. D., Klotz, L. C., and Fresco, J. R. (1964). *C. R. Acad. Sci. Paris*, **259**, 3104–7.

17. Neumann, E. and Katchalsky, A. (1970). *Ber Bunsenges.*, **74**, 868–79.

18. Thrierr, J. C. and Leng, M. (1972). *Biochim. Biophys. Acta*, **272**, 238–51.

19. Arnott, S. and Bond, P. J. (1973). *Nature N. B.*, **244**, 99–101; Arnott, S. et al. (1973). *J. Mol. Biol.*, **81**, 107–22.

20. Arnott, S. (1970). *Prog. Biophys.*, **21**, 265–319.

21. Arnott, S. and Hukins, D. W. L. (1973). *J. Mol. Biol.*, **81**, 93–105.

22. Ishikawa, F. et al. (1972). *J. Mol. Biol.*, **70**, 475–90.

23. Bernardi, G. and Timasheff, S. N. (1970). *J. Mol. Biol.*, **48**, 43–52.

24. Sarocchi, M. Th. and Guschlbauer, W. (1973). *Eur. J. Biochem.*, **34**, 232–40.

25. Baldwin, R. L. (1971). *Acc. Chem. Res.*, **4**, 265–72.

26. Record, M. T. (1967). *Biopolymers*, **5**, 43–52.

27. Arnott, S. et al. (1974). *J. Mol. Biol.*, **88**, 509–21; 523–33; 551–52.

28. Thiele, D. and Guschlbauer, W. (1969). *Biopolymers*, **8**, 361–78; (1971). *ibid.*, **10**, 143–57; Thiele, D., Guschlbauer, W., and Favre, A. (1972). *Biochim. Biophys. Acta*, **272**, 22–26.

29. Thiele, D. et al. (1973). *Mol. Biol. Rep.*, **1**, 149–54; 155–60.

30. Bachner, L., and Massoulié, J. (1973). *Eur. J. Biochem.*, **35**, 95–105.

31. Byrd, C., Ohtsuka, E., Moon, M. W., and Khorana, H. G. (1965). *Proc. Nat'l. Acad. Sci., U.S.A.*, **53**, 79–86.

32. Wells, R. D., Ohtsuka, E., and Khorana, H. G. (1965). *J. Mol. Biol.*, **14**, 221–40.

33. Wells, R. D. et al. (1967). *J. Mol. Biol.*, **27**, 237–64; 265–86; (1970). *ibid.*, **54**, 465–97.

34. Bram, S. (1971). *Nature, N. B.*, **232**, 174–76.

35. Ts'o, P. O. P., Rappaport, S. A., and Bollum, F. J. (1966). *Biochemistry*, **5**, 4153–60.

36. Bobst, A. M., Rottmann, F., and Cerutti, P. (1969). *J. Mol. Biol.*, **46**, 221–34.

37. Zmudzka, R. and Shugar, D. (1971). *Acta Biochim. Pol.*, **18**, 321–37.

38. Eigen, M. and Pörschke, D. (1970). *J. Mol. Biol.*, **53**, 123–43.

39. Delisi, C. and Crothers, C. (1971). *Biopolymers*, **10**, 2323–43.

40. Lezius, A. and Dohmin, E. (1973). *Nature, N. B.*, **244**, 169–70.

41. Lomant, A. J. and Fresco, J. R. (1973). *Biopolymers*, **12**, 1889–1903.

42. Michelson, A. M. (1962). *Biochim. Biophys. Acta,* **55,** 841–48.

43. Brahms, J., Michelson, A. M., and Van Holde, K. E. (1966). *J. Mol. Biol.,* **15,** 467–85; Brahms, J., Maurizot, J. C., and Michelson, A. M. (1967). *J. Mol. Biol.,* **25,** 465–80; 481–93.

44. Bush, C. A. and Tinoco, I. (1967). *J. Mol. Biol.,* **23,** 601–10; Tinoco, I. (1968). *J. Chim. Phys.* (Paris), **65,** 90–97.

45. Guschlbauer, W., Frič, I., and Holý, A. (1972). *Eur. J. Biochem.,* **31,** 1–13.

46. Ts'o, P. O. P. *et al.* (1969). *Biochemistry,* **8,** 997–1029.

47. Kondo, N. S. and Danyluk, S. S. (1972). *J. Amer. Chem. Soc.,* **9,** 5121–22.

48. Brahms, J., Maurizot, J. C., and Pilet, J. (1969). *Biochim. Biophys. Acta,* **186,** 110–23.

49. Solie, T. N. and Schellmann, J. A. (1968). *J. Mol. Biol.,* **33,** 61–77.

50. Schweitzer, M. P. *et al.* (1968). *J. Amer. Chem. Soc.,* **90,** 1042–55.

51. Ts'o, P. O. P. and Chan, S. I. (1964). *J. Amer. Chem. Soc.,* **86,** 4176–81.

52. Kyogoku, Y., Lord, R. C., and Rich, A. (1967). *J. Amer. Chem. Soc.,* **89,** 496–505; Thomas, G. J. and Kyogoku, Y. (1967). *J. Amer. Chem. Soc.,* **89,** 4170–78.

53. Gratzer, W. B. and McClare, C. W. F. (1967). *J. Amer. Chem. Soc.,* **89,** 4224–25.

54. Hoogsteen, K. (1963). *Acta Crystallog.,* **16,** 907–15.

55. Haschemeyer, A. E. V. and Sobell, H. M. (1965). *Acta. Crystallog.,* **18,** 525–32.

56. Sobell, H. M. (1972). *Jerusalem Symp.,* **4,** 124–48.

57. Haschemeyer, A. E. V. and Sobell, H. M. (1965). *Acta Crystallog.,* **19,** 125–30.

7 Errors and Mutations

In Chapter 3 (Section 3.2.2), the reactions of nucleic acid bases with chemical reagents were presented. Similar reactions are widely used to modify nucleic acids chemically, but are generally less efficient because of their secondary structure.

Probably the most widely used mutagen is nitrous acid. It reacts with amino groups and replaces them by OH groups, which, in turn, rearrange to the corresponding keto compounds: A→I, C→U, G→X. Such transitions (1) in Ade or Cyt will change the base pairs from A-T to I-T→I-C→ G-C and G-C to G-U→A-U→A-T respectively, whereas deamination of Gua yields G-C to X-C→X-T→A-T.

Nitrous acid-induced mutations in tobacco mosaic virus (TMV) have contributed to the derivation of the genetic code (see Table 8.1) through amino acid analyses determining replacements in the coat protein (2). Since only C→U, G→A or A→G transitions are possible in the case of a nitrous acid point mutation, the number of possible replacements for a given amino acid is small. Thus, an Ala→Val transition corresponds to a change of the codon GCN to GUN, whereas an Ala→Thr transition must be due to a change in the first letter: GCN→ACN (See Chapter 8, Table 8.1).

Another widely used mutagen is hydroxylamine, NH_2OH. Since it reacts almost exclusively with Cyt to yield Ura, one obtains another transition G-C→G-U→A-U→A-T.

Formaldehyde and alkylating agents, and especially bifunctional ones (but also certain antibiotics like mitomycin), are highly mutagenic, since they induce crosslinks which are mainly on Guo. These crosslinks prevent DNA from unwinding and thus block DNA synthesis. Such cross links may also be the origin of deletions of certain control regions of DNA and, thus, the cause of malignancies. This may explain the fact that certain alkylating agents are carcinogenic.

7.2
Base Analogues

Similar transitions in base pairs can be observed with base analogues. Such analogues as 5-bromo-Ura (br⁵Ura,BU), 2-amino-purine (n²Pur,AP), or 2,6-diamino-purine (n²n⁶Pur, DAP), can be incorporated into DNA (3) instead of Thy, Ade, or Gua, respectively. Upon replication, they can induce mutations, either by changing to the enol form (BU) or imino form (AP) or by an ambiguity of base pairing (see Fig. 7.1). The result is an A-T to G-C shift, quite similar to that observed with chemical mutagens. Furthermore, bromouridine or iodouridine mutants are more sensitive to UV and X-ray radiation. This may be due to the proximity of the negative phosphate groups to the electronegative halogen atom. Also, the fact that the halogeno-Urd's absorb at higher wavelengths than Thd renders them susceptible to photochemical reactions at higher wavelengths. These reactions sometimes involve dehalogenation, with free radical formation and subsequent strand breaks.

FIGURE 7.1. Mutagenic effect of 2-aminopurine (a) and 5-bromouracil (b) upon incorporation into DNA. Note the tautomeric form in the third base pair.

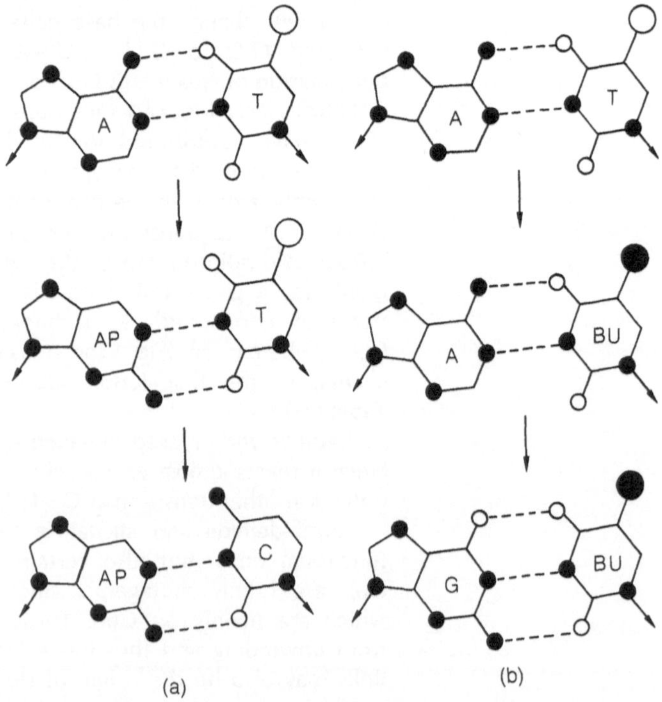

(a)   (b)

A very intense search for antiviral and cancerostatic agents has led to the synthesis of countless nucleoside analogues. Virtually all possible chemical modifications (on the bases, sugars, and even the phosphate) have been made and tested *in vitro* and *in vivo*. Two of the most intensively studied antimetabolites are arabinosyl cytosine (araCyd) and formycin A (see Fig. 3-1). The former is used in the clinical treatment of oncogenic viruses and leukemia. After transformation to the triphosphate it is incorporated into DNA, where it may be a chain terminator. Formycin A and B (A is the Ado analogue, B the Ino analogue) are both in the *syn* conformation and have strong antiviral activity.

fldUrd, brdUrd, and iodUrd are incorporated into DNA and probably block mRNA transcription. They have proved to be effective against virus infections, when given in the form of a free base.

But 5-Fluoro-uracil (flUra) is preferentially incorporated into RNA after conversion to the riboside and phosphorylation. The size of fluorine makes flUra more of an Ura analogue, whereas the large halogens simulate the size of the methyl group of Thy (see Fig. 7.1). The high electronegativity of fluorine, however, completely changes its electronic structure and properties, and flUra is a powerful inhibitor of rRNA synthesis. flUra, by the bias of inhibiting thymidylate synthesis, also blocks DNA synthesis.

Further information on nucleoside analogues may be found in a short, but very comprehensive review (4), which includes an extensive bibliography. A monograph by Roy-Burmann (5) gives a very good account.

7.3
Antibiotics,
Pigments, Dyes

Several kinds of dye–nucleic acid interactions are found. Since most dyes are planar aromatic (frequently positively charged) compounds, they (a) can intercalate (6), i.e., slide between two successive base pairs into the helix, thus extending its length; (b) can bind at low concentrations one by-one to the negative phosphates of the DNA; (c) can connect two phosphates when doubly charged; and (d) at high concentrations, can cooperatively stack along the phosphates. Here, we shall be concerned mainly with intercalation.

Most of the attention has been focused on the three dyes: actinomycin D, ethidium bromide, and acridine orange (Fig. 7.2). All three intercalate. Actinomycin D forms strong (7) specific complexes with dGuo in the free form or in DNA. The crystal structure of the actinomycin-dGuo complex has been demonstrated by X-ray diffraction (8). This structure (Fig. 7.3) has an internal twofold axis in the phenoxazone ring plane. Each peptide ring forms a bond with the amino group of the upper and lower dGuo. It is of interest that the two dGuo molecules of this structure are

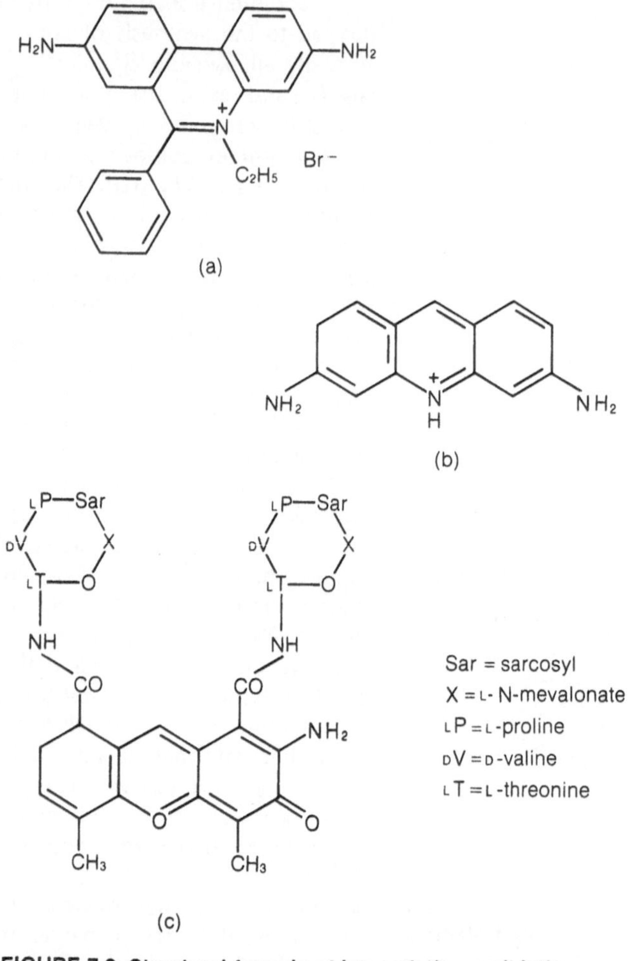

(a)

(b)

Sar = sarcosyl
X = L- N-mevalonate
L P = L-proline
D V = D-valine
L T = L-threonine

(c)

FIGURE 7.2. Structural formula of intercalating antibiotics and dyes. (a) Ethidium bromide; (b) Proflavin; (c) Actinomycin D.

in the exact positions as they are in the Watson–Crick poly(dG-dC) helix [Fig. 7.3(c)]. Theories on the possible model role of this complex for repressor binding (which must have a twofold axis since it is a tetramer) have been formulated (8).

Ethidium bromide—and other intercalating dyes—have been widely used (9–12) to measure supercoiling in closed DNA circles. Supercoils will be released either by single strand breaks or by intercalating dyes to uncoil the DNA helix. Since circles have lower sedimentation coefficients than supercoiled circles, they can be titrated (Fig. 7.4). The minimum of the sedimentation constant vs. concentration curves corresponds to the point where no supercoil exists.

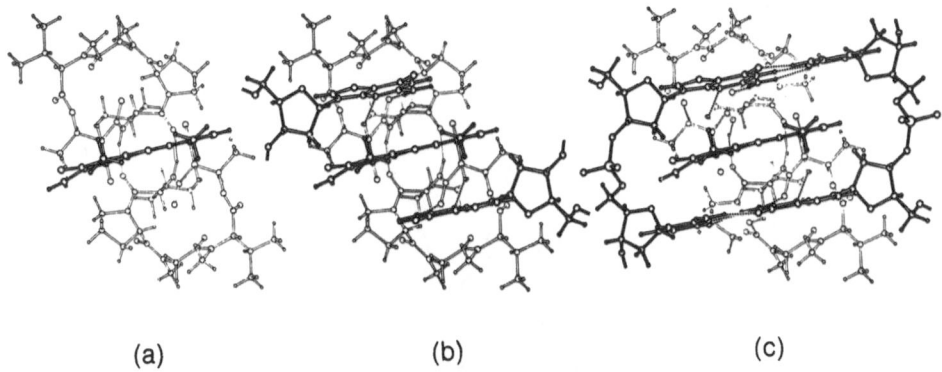

(a)   (b)   (c)

FIGURE 7.3. Stereochemical model of actinomycin D (a), its complex with dGuo (b), and the complex actinomycin [d(GpC)]·[d(GpC)](c). (From Ref. 8.)

Additional intercalation will overwind the helix in the opposite direction, and the sedimentation constant will again increase. From the number of dye molecules bound, one can determine the number of extra turns in the supercoil (see Fig. 5.10).

Acridine dyes (acridine orange, proflavine) (Fig. 7.2) will produce so-called frame-shift mutants. In this case, the intercalated dye mimics an additional base pair and, during replication, the daughter strands will contain an extra base at this point. By using carefully chosen point mutants, Crick et al. (13) proved the three-letter genetic code.

7.4 Mutagenesis by Radiation (UV, X-Rays)

X-ray irradiation (14) of cells attacks DNA directly or indirectly. Direct effects are double strand breaks, which are always lethal. Indirect effects are due to peroxides or radicals formed in the surrounding medium by the radiation. These compounds can then attack DNA, with various processes possible. Single strand breaks in the phosphodiester backbone are generally non-lethal, since DNA ligase (repair enzyme) can easily close the break. Base damage is frequently more serious and can cause mutations. These effects may be seen at the level of replication, transcription, or translation. Most peroxides and radicals can be trapped by so-called scavengers, such as histidine or sulfhydryl compounds, which thus serve to attenuate radiation damage. The effect of gamma rays are similar to that of X-rays.

Ultraviolet rays (15), which are much less energetic than X-rays, mainly affect the C^5-C^6 double bonds of the pyrimidines, forming hydrates (reversible) or dimers (see also Chapter 3, Section 3.2.2.). Dimers, if in the same strand (i.e., on neighboring Thy residues), are eliminated by nucleases,

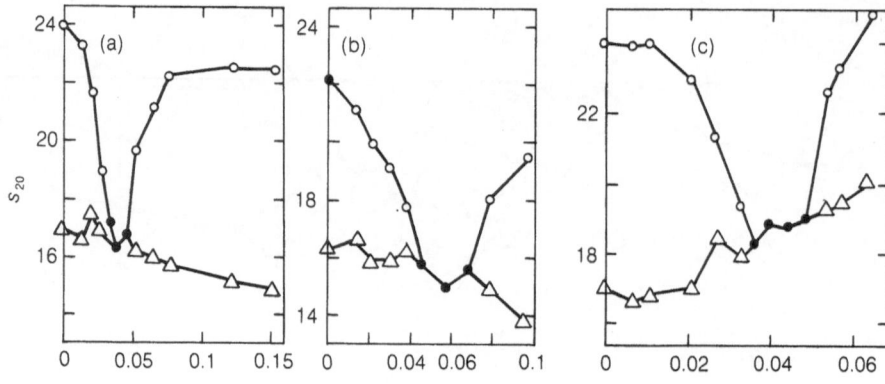

FIGURE 7.4. Effect of intercalation on the sedimentation coefficient of ϕX 174 replicative form. (a) Ethidium bromide (72%; closed circles); (b) proflavin (67%; closed circles); (c) actinomycin (75%; closed circles). Open circles: replicative form; open triangles: nicked DNA; Hr: dye bound/nucleotide. (From Ref. 10.)

and the gap is closed by repair enzymes. Dimers on opposite strands are much more serious, since they produce cross links that are generally lethal mutations (16).

References

1. Freese, E. (1959). *Proc. Nat'l. Acad. Sci., U.S.A.,* **45,** 622–31.
2. Wittmann, H. G. (1961). *Naturwissen.* **48,** 729–34; (1963). ibid., **50,** 76–89; Wittman, H. G. and Wittmann-Liebold, B. (1963). *Cold Spring Harb. Symp. Quart. Biol.,* **28,** 589–95.
3. Freese, E. (1959). *J. Mol. Biol.,* **1,** 87–105.
4. Shugar, D. (1974). *FEBS Letters, Suppl.,* **40,** S48–62.
5. Roy-Burman, P. (1970). *Analogues of Nucleic Acids Components.* New York: Springer-Verlag.
6. Lerman, L. S. (1961). *J. Mol. Biol.,* **3,** 18–30.
7. Wells, R. D. and Larson, J. (1970). *J. Mol. Biol.,* **49,** 319–42.
8. Sobell, H. M. (1973). *Prog. Nucleic Acid Res. Mol. Biol.,* **13,** 153–90.
9. Crawford, L. V. and Waring, M. (1967). *J. Mol. Biol.,* **25,** 23–30.
10. Waring, M. (1970). *J. Mol. Biol.,* **54,** 247–79.
11. Bauer, W. and Vinograd, J. (1968). *J. Mol. Biol.,* **33,** 141–71.
12. Bauer, W. and Vinograd, J. (1970). *J. Mol. Biol.,* **47,** 419–35; **54,** 281–98.
13. Crick, F. H. C. *et al.* (1961). *Nature,* **192,** 1227–32.
14. Kanazir, D. T. (1969). *Prog. Nucleic Acid Res. Mol. Biol.,* **9,** 117–222.
15. Weiss, J. J. (1963). *Prog. Nucleic Acid Res. Mol. Biol.,* **3,** 103–40.
16. Wacker, A. (1961). *Prog. Nucleic Acad. Res. Mol. Biol.,* **1,** 369–99.

8 The Structure of Ribonucleic Acids

8.1 Protein Biosynthesis

We shall not attempt to discuss protein synthesis at any length, only as much as is needed to consider some structural aspects.

In the process of protein biosynthesis, genetic information is translated from the nucleotide sequence of the mRNA into proteins. Since there may be only one copy in the DNA of the gene for a given enzyme, from which thousands of molecules of the enzyme must be synthesized, an amplification mechanism, which must be rapid and exact, must function during translation. This mechanism is now understood and involves principally three different kinds of RNA: messenger RNA (mRNA), ribosomal RNA (rRNA), and transfer RNA (tRNA). The existence of mRNA was postulated (1) in 1960 and shortly afterwards (2,3) demonstrated experimentally. It was shown to be a single-stranded (and thus relatively fragile) RNA of varying length. Its turnover was found to be rapid in bacteria. Ribosomes, on the contrary, are relatively stable due to their protein coat. They are the main RNA-containing constituent of the cytoplasm and are found in the endoplasmic reticulum (4). Their existence has been known for a long time. Crick (5) proposed the existence of tRNA, the adaptor molecule, which was identified as the earlier described sRNA (6). Its distinctive properties (esterification with amino acids, unusual bases, and low molecular weight) made it an ideal compound for study by chemical, enzymatic, and physicochemical techniques.

These three elements assemble during protein biosynthesis (Fig. 8.1) according to a very precise mechanism (7). The 70 S ribosomes have two subunits; the messenger is bound to the 30 S subunit, while tRNA is bound to the 50 S subunit. Three bases of tRNA act as the anticodon, pair-

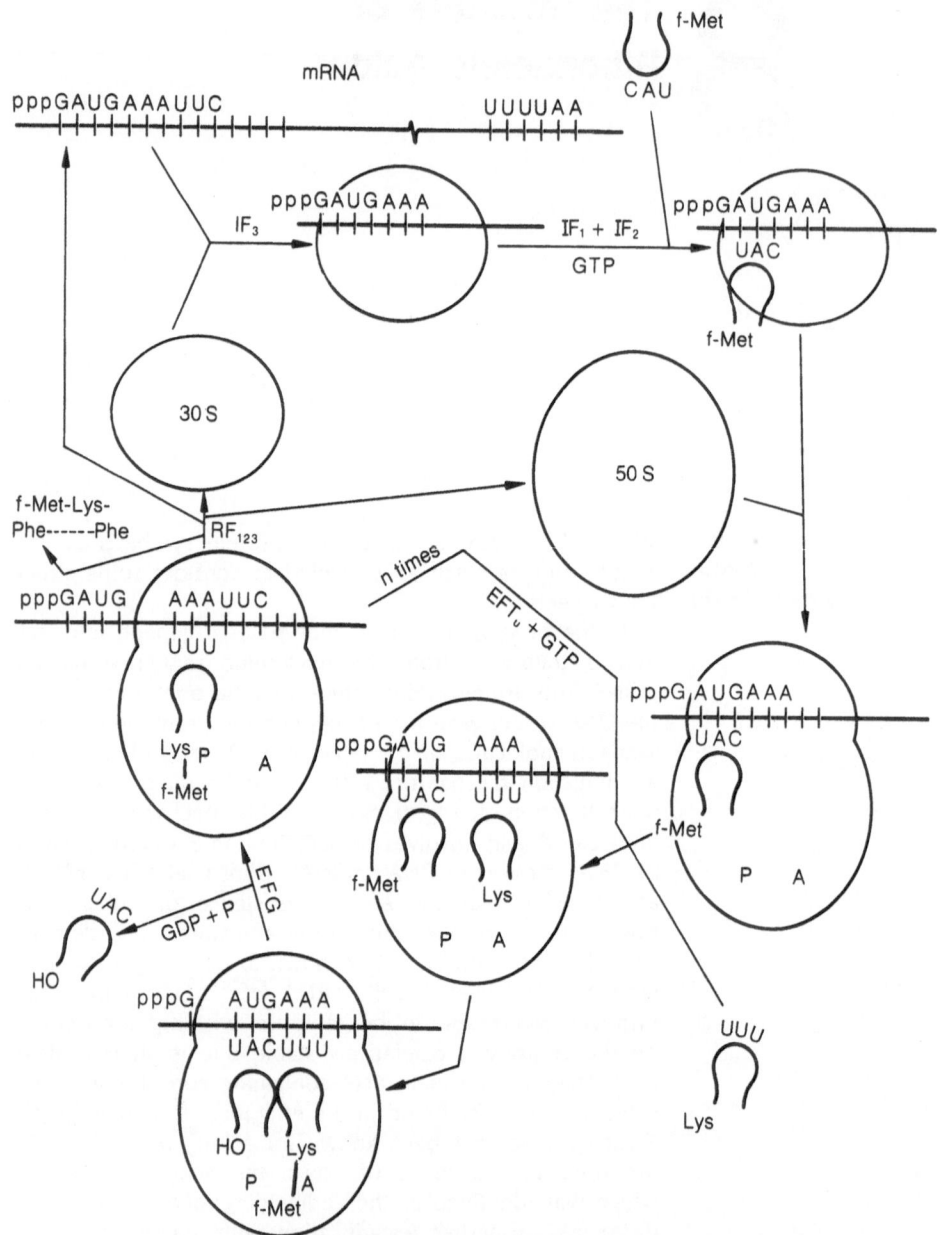

FIGURE 8.1. Scheme of protein biosynthesis in *E. coli*. Initiation factors (IF), elongation factors (EF), release (termination) factors (RF).

ing specifically with the messenger codon, which is defined by the genetic code (Table 8.1). The observation that the third base of the anticodon is a modified base—generally hypoxanthine—which does not bind a *priori* with the corresponding codon base, led Crick (8) to suggest the existence of "wobble pairs," (Fig. 8.2), which would be slightly out of the usual base pairing geometry. Evidently these three base pairs alone are not sufficient to stabilize the tRNA with the mRNA. Therefore, another binding site was postulated and eventually found on the 50 S subunit.

Initiation of the polypeptide chain always occurs with an N-formylmethionyl-tRNA, for which a specific tRNA species exists; it is bound to the 30 S subunit of the ribosome. Met-tRNA ligase can acylate two met-tRNA's. Only one of these, however, can be formylated to the initiator-tRNA. The initiation complex associates with the 50 S subunit (Fig. 8.1) where the f-met-tRNA is positioned on a precise site (P-site) of the ribosome.

Translation takes place by a sequential passing of the mRNA in the 5'→3' direction, which places the tRNA in the A-site of the ribosome, while a new tRNA is positioned in the P-site. Peptide bond formation is accomplished in this configuration. As many protein chains can be assembled as there are ribosomes on the messenger strand (Fig. 8.3). (There are some 10,000 ribosomes in a bacterial cell.) These polyribosomes (or polysomes) permit the rapid ampli-

FIGURE 8.2. Wobble pairs according to Crick (Ref. 8).

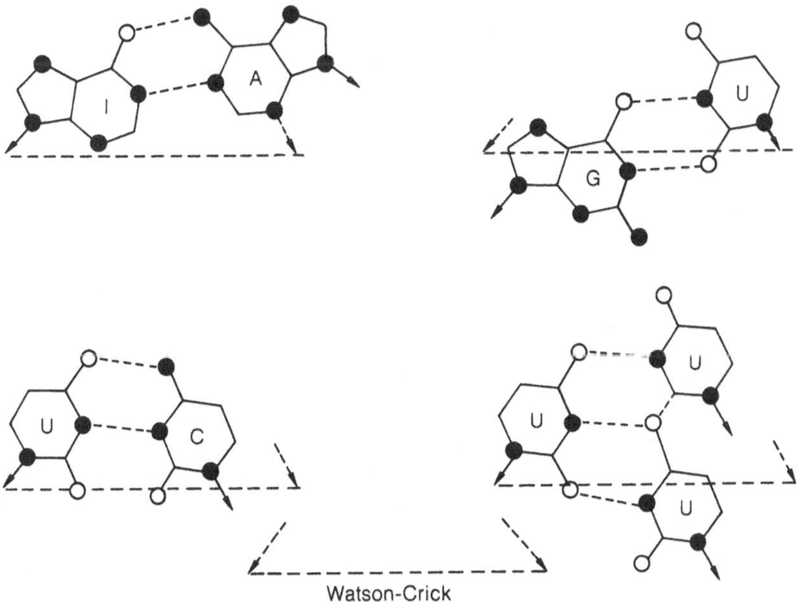

Watson-Crick

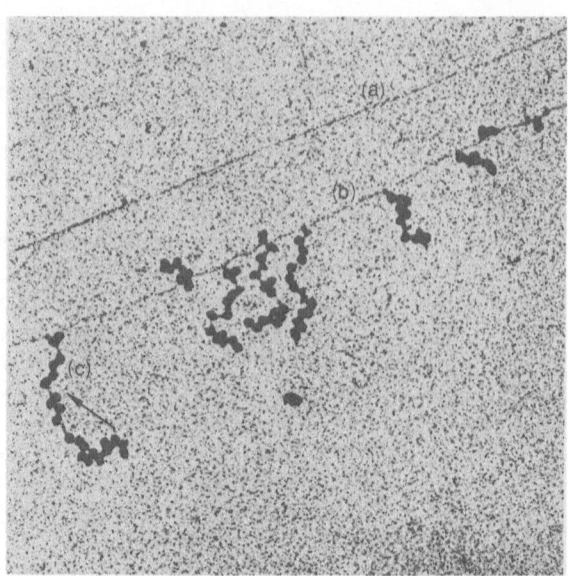

FIGURE 8.3. Electron micrograph of *E. coli* chromosomes
(a) Inactive chromosome fragment; (b) active
chromosome during transcription. Small dots on the
chromosome where the polysomes (c) are attached are
RNA polymerase molecules, which transcribe the DNA
into RNA from right to left. The ribosomes attach
themselves to the transcribed mRNA, which peels off
during transcription. The ribosomes are transcribed on
other parts of the chromosome. The polysomes (c) thus
formed start translating the mRNA (protein synthesis) in
the direction from the dangling end of the polysome
toward the chromosomal attachment point [direction of
the arrow near (c)]. Translated protein chains cannot
be seen.

fication of the genetic message into proteins. On the poly-
ribosome, each tRNA charged with an amino acid adds its
amino acid to the nascent polypeptide chain. After this
reaction, the free tRNA is returned to the pool. All these
steps use a series of protein factors and GTP and take place
at the A and P sites of the ribosome.

Chain termination, on the other hand, is signaled by the
codons UAG, UAA, or UGA, to which normally no tRNA
corresponds; only in suppressor mutants, do certain tRNA's
(like su$^-$-tyr-tRNA) permit the reading of termination signals.
The 70 S ribosome dissociates into 50 S and 30 S subunits
upon chain termination.

8.2
Transfer RNA

Transfer RNA (tRNA) plays a key role in the translation process in the amino acid activation reaction

$$aa_x + ATP + Enz_x \rightleftharpoons (AMP\text{-}aa_x\text{-}Enz_x) + PP_i$$
$$(AMP\text{-}aa_x\text{-}Enz_x) + tRNA_x \rightleftharpoons aa_x\text{-}tRNA_x + Enz_x + AMP$$

The carboxyl group of the amino acid is esterified with the 3'OH group of the terminal Ado of the tRNA in the presence of Mg. On the ribosome complex, through the action of the elongation factors, GTP and the peptide synthetase, the peptide bond is formed. This implies that breakage of the ester bond of the second-to-last aa-tRNA occurs, and its carboxyl group reacts with the amino function of the last aa-tRNA.

Because tRNA is involved in all these processes of protein synthesis, it therefore must have several distinct recognition sites for various enzymes and polymers. At present, only one of these sites is unequivocally known: the anticodon, by which tRNA interacts with mRNA. Although many publications have appeared on the different functions of tRNA, no conclusive evidence is yet available to associate other regions of the tRNA molecule with particular functions.

8.2.1 CHEMICAL STUDIES ON tRNA IN SOLUTION

Ever since tRNA was identified, it was clear that it was an unusual type of RNA (9). Its molecular weight of about 25,000, its rather low axial ratio, the modified bases, its resistance to nucleases (10), and the limited splitting at unique spots (which eventually permitted the elucidation of its sequence), the ubiquitous -C-C-A terminal sequence, which could be reversibly removed, all indicated that tRNA had unique properties compared to other RNA's (9).

This brings us to the multiplicity of tRNA's. It was recognized early that not only are all the different tRNA's very similar in chemical and physicochemical behavior (which is not too surprising), but also that, in a given organism, more than one tRNA existed for each amino acid, and coded with different codons (Table 8.1). This finding made the development of rather sophisticated separation techniques necessary. A widely used technique is countercurrent distribution. It is based on Nernst's law. For two nonmiscible solvents, any solute has a characteristic partition coefficient (i.e., the ratio of activities in the two solvents at equilibrium). If two tRNA's have different partition coefficients in a given solvent pair, one can be enriched in the first solvent and the other in the second. Successive extractions performed by renewing one or both solvents effect the separation (Fig. 8.4). Countercurrent distribution is performed in machines containing hundreds of mixing and separation tubes.

Table 8.1   The Genetic Code

First letter	Second Letter								Third letter
	U		C		A		G		
U	UUU	Phe	UCU	Ser	UAU	Tyr	UGU	Cys	U
	UUC	Phe	UCC	Ser	UAC	Tyr	UGC	Cys	C
	UUA	Leu	UCA	Ser	UAA	Ochre[a]	UGA	Azur[a]	A
	UUG	Leu	UCG	Ser	UAG	Amber[a]	UGG	Trp	G
C	CUU	Leu	CCU	Pro	CAU	His	CGU	Arg	U
	CUC	Leu	CCC	Pro	CAC	His	CGC	Arg	C
	CUA	Leu	CCA	Pro	CAA	Glu	CGA	Arg	A
	CUG	Leu	CCG	Pro	CAG	Glu	CGG	Arg	G
A	AUU	Ile	ACU	Thr	AAU	Asp	AGU	Ser	U
	AUC	Ile	ACC	Thr	AAC	Asp	AGC	Ser	C
	AUA	Ile	ACA	Thr	AAA	Lys	AGA	Arg	A
	AUG	Met[b]	ACG	Thr	AAG	Lys	AGG	Arg	G
G	GUU	Val	GCU	Ala	GAU	Asp	GGU	Gly	U
	GUC	Val	GCC	Ala	GAC	Asp	GGC	Gly	C
	GUA	Val	GCA	Ala	GAA	Glu	GGA	Gly	A
	GUG	Val	GCG	Ala	GAG	Glu	GGG	Gly	G

[a] Termination codons.

[b] Also initiation codon F-Met.

As soon as it became apparent that all tRNA sequences could be assembled in a cloverleaf (12,13) (Fig. 8.5), a number of studies were performed to define the tertiary structure of tRNA (14,15). A proliferation of models for tRNA appeared, more or less well incorporating the chemical data known at the time. It is this evidence that will now be evaluated.

All tRNA sequences (Fig. 8.5) known so far have, in addition to the cloverleaf structure, several features in common. The three loops I, II, and IV are always present. Loop I, which varies in size from one species to another, always contains one or several dihydro-Urd (D) residues. This non-aromatic base is also present in loop III in some tRNA's. Loop II contains the anticodon, flanked on the 3'-side by a substituted purine; some of these substituents may be long aliphatic chains, as in the cytokinins (16). Loop IV always contains the characteristic -T-Ψ-C-G sequence which is probably necessary for binding of tRNA to the ribosome. On the other hand, large variations in the four helical regions are observed for different tRNA's. The -C-C-A terminus, where the amino acid is esterified, is always present.

Although certain bases and groups are in loop regions (Fig. 8.5) and not in helical regions, the T-Ψ-C-G loop appears to be protected against chemical attack by acrylo-

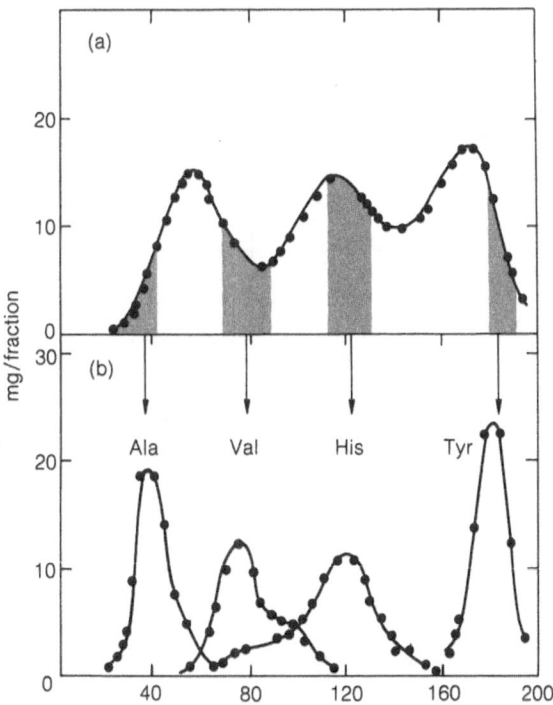

FIGURE 8.4. Countercurrent distribution of yeast tRNA.
(a) First run; (b) second run of the tubes in the
shaded areas of the top part of the graph.

nitrile or carbodiimide. On the other hand, the anticodon
loop is readily attacked by chemical reagents, as is the
D-loop I. Highly selective reagents have yielded valuable
information on the exposure of certain bases. Kethoxal
(CH_3-CH(OEt)-C$=$O-CHO), which reacts only with the
amino group of unpaired Gua residues, attacks only two
Guo in tRNA, those in the anticodon loop and in the
D-loop I. Similarly, perphthalic acid, which forms N^1-oxides
with free Ado residues, will only react with those in these
two loops (Fig. 8.6).

One of the most characteristic and informative reactions
was the photochemical bridging of s^4Urd (which replaces
Urd in position 8 in some tRNA's) with the Cyt in position
13 (17). Since this reaction, which is similar to the Thd
dimerization (see Chapter 3, Section 3.2.2, p. 25, forms
covalent bonds between the two molecules, they must be
in close proximity. Particularly interesting is the fact that

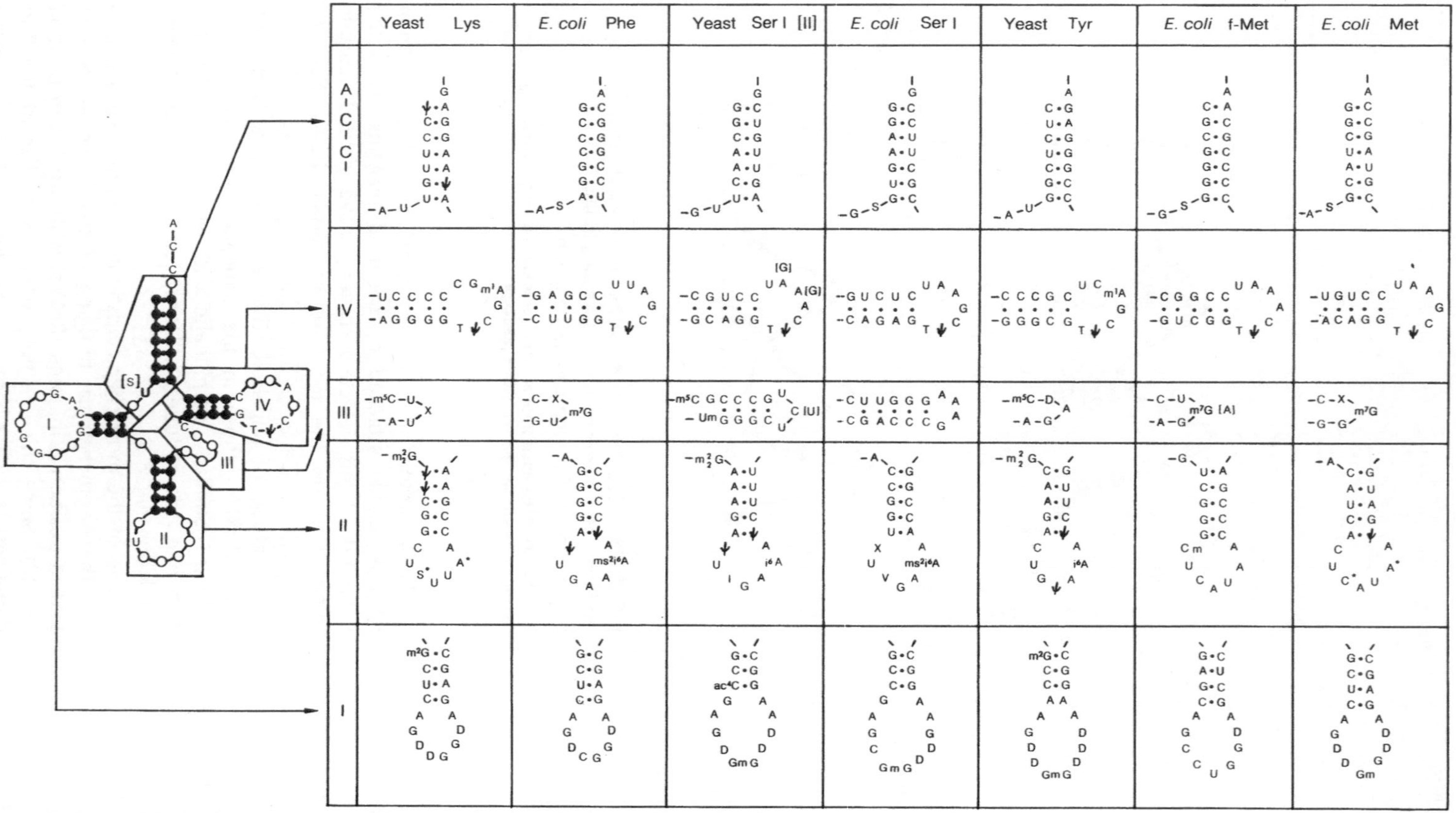

FIGURE 8.5. Sequences of various tRNA's from *E. coli* and yeast. Sequences have been arranged in such a manner that the similarities and differences will be most apparent. Carbamoyl-threonyl-A (A*), modified bases (S*, C*, A*), details unknown; unknown base (X), uracil-5-acetic acid (V), 2-thiomethyl-*N*⁶-isopentenyladenine (ms²i⁶A), *N*⁶-isopentenyladenine (i⁶A).

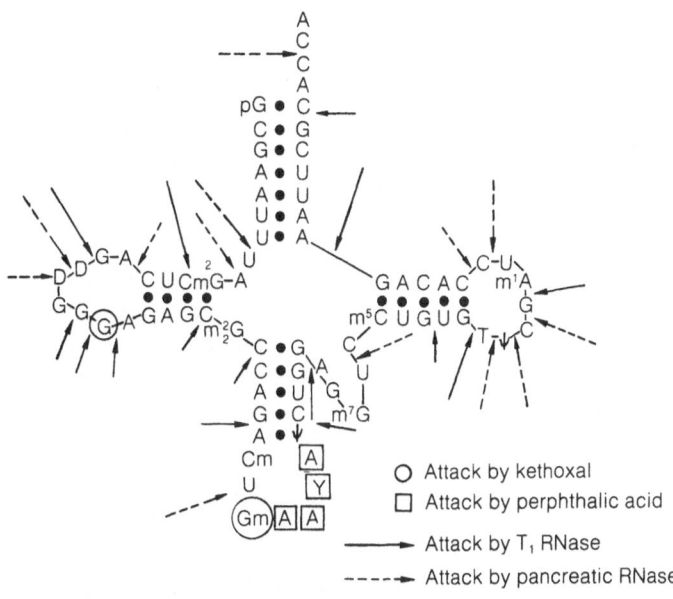

FIGURE 8.6. Enzymatic attack of yeast tRNA$_{Phe}$. (After Ref. 14.)

tRNA, thus linked, retains virtually all its biochemical properties, although the rate of some of its reactions (e.g., acylation) is lower.

The chemical evidence thus suggests that tRNA is a very compact molecule in which two of the loops (I and II) are reasonably accessible, whereas loops III and IV (the T-Ψ-C-G loop) are protected and folded in such a way that they are not open to attack. Furthermore, residues 8 (which is always a Urd or s⁴Urd) and 13 (which is always a Cyd) must be very close to each other.

Cyanoethylation of the Urd residue takes place in loop IV and does not inhibit acylation of the tRNA, although the rate is lower.

8.2.2 ENZYMATIC STUDIES ON tRNA IN SOLUTION

Limited attack by nucleases at 0° shows that certain regions of tRNA are still accessible. PNPase (Chapter 3, Section 3.4.2, p. 27) acts sequentially on the 3'OH end of unstructured polynucleotides. But tRNA is only slightly affected by polynucleotide phosphorylase (18), and the decrease in acylation capacity parallels the extent of phosphorolysis. This indicates that some of the tRNA molecules are not attacked at all (since they still accept amino acids) and that the molecules attacked by the enzyme are completely degraded. Therefore, the -C-C-A terminus must be protected. Heating increases the enzymatic availability of the

-C-C-A terminus (Fig. 8.7). Similarly, the 5'-phosphate is inaccessible to phosphatase action.

Endonucleases, like pancreatic RNase or takadiesterase (Chapter 3, Section 3.4.2, p. 27) split tRNA to any extent mainly in the loops (10,14,19) (Fig. 8.6). This reaction has been used to split tRNA molecules into halves and quarters, which could be isolated, characterized, and then further degraded. By this means, the nucleotide sequences of many tRNA's have been elucidated (12–14). Furthermore, the reassembly of such fragments has been used to determine which part of the tRNA molecule is necessary for which function. From such studies (19), it was concluded that the recognition by the tRNA ligase (aminoacyl-tRNA-synthetase) takes place in the stem region, and the loops and other helical regions are not involved (Fig. 8.8).

The rare bases of tRNA present an interesting case of enzymatic modification. After the biosynthesis of tRNA on the DNA chain, where only the four natural bases Ade, Gua, Cyt, and Ura are incorporated, some two dozen enzymes act specifically to modify some of the tRNA bases. Some are modified on the unfolded tRNA chain, whereas others, like the methylated bases, are formed only after the tRNA has assumed its tertiary structure. It is thus not surprising to find these methylated bases in the loops. Some of these bases, such as cytokinins (e.g., isopentenyl-Ade) (16) in the anticodon loop, have long aliphatic chains, which take up considerable space.

FIGURE 8.7. Extent of phosphorolysis of total tRNA as a function of temperature. (From Ref. 18.)

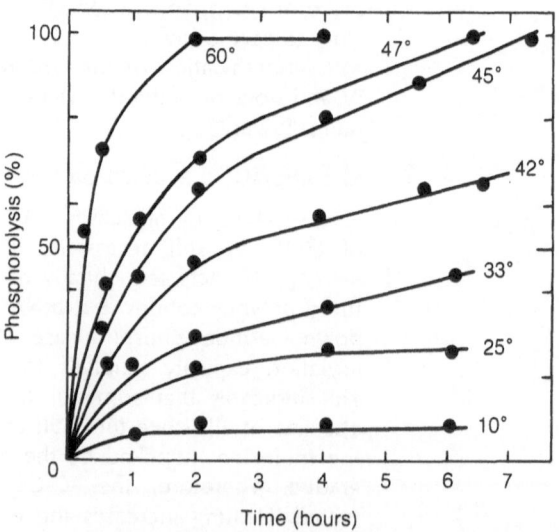

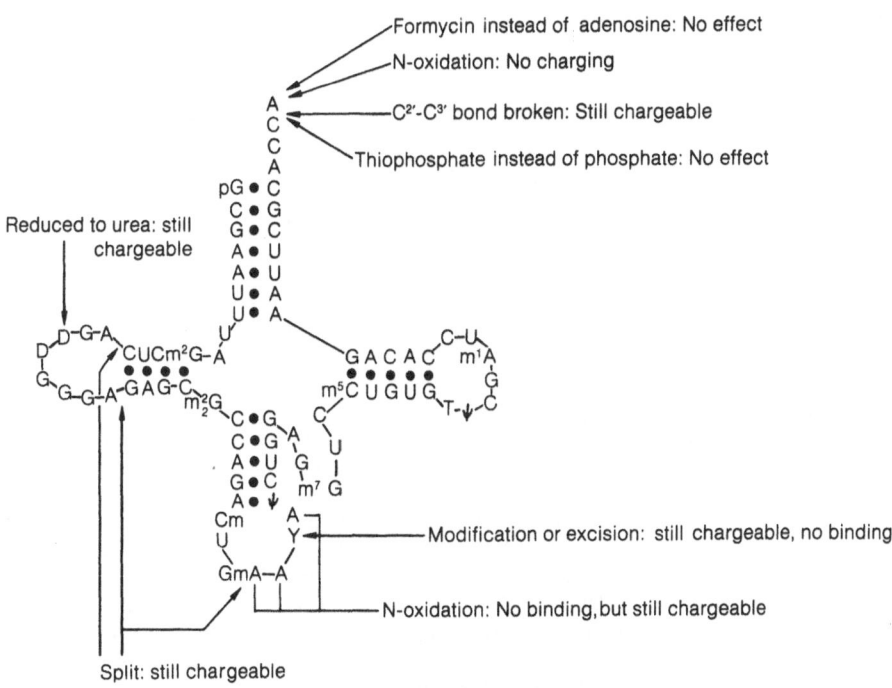

FIGURE 8.8. Influence of modifications on biosynthetic properties of tRNA$_{Phe}$. (After Ref. 14.)

8.2.3 PHYSICOCHEMICAL STUDIES ON tRNA IN SOLUTION

It was recognized very early, by the use of spectroscopic techniques (11), that tRNA had a tertiary structure, with different regions. In fact tRNA showed a "biphasic" melting profile (Fig. 8.9), which could be analyzed in terms of regions with different G-C content. In the presence of bivalent ions, the secondary and tertiary structure remains stable over a rather large temperature range (Fig. 8.9), as evidenced by spectroscopic and hydrodynamic properties. Above 60°, however, rapid structural disintegration was observed (20) and the amino acid charging capacity dropped concominantly.

That two divalent cations are tightly bound to native tRNA has been established by NMR studies. If these cations are eliminated, tRNA assumes another structure, which is, however, not biologically active; it can be renatured at 60° in the presence of Mg ions. Recently, Kearns et al. (21) measured the NMR spectra of the hydrogen-bonded protons of the helical regions of leucyl tRNA. An exact assignment of an NMR peak for each proton of each base could be made. The ring current of the neighboring base pairs influences the chemical shifts of the N-H protons. With a number of known tRNA sequences as reference, the method

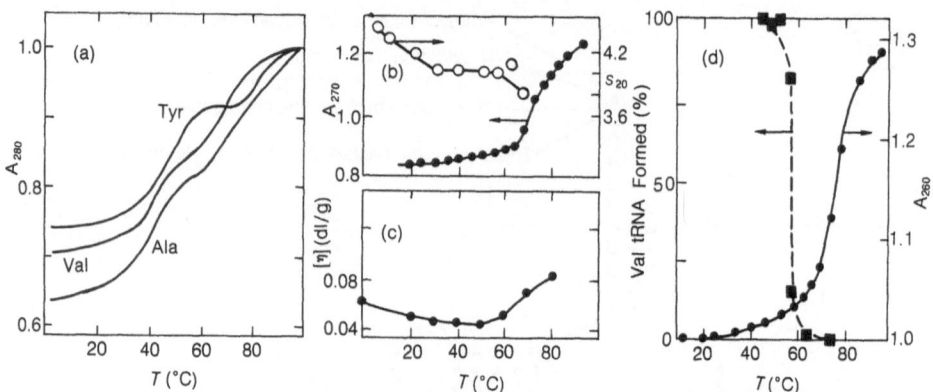

FIGURE 8.9. Comparison of various properties of tRNA's as a function of temperature. (a) Optical melting curves of three purified yeast tRNA's in 0.15 M NaCl, pH=7; (b) temperature dependence of absorption (•) and sedimentation coefficient (o) (at infinite dilution); and (c) intrinsic viscosity of yeast tRNA (unfractionated) in 0.15 M KCl, 0.01 M potassium cacodylate, 0.005 M MgCl$_2$, and 0.0005 M EDTA; (d) valine acceptor activity (squares) and absorption (circles) of unfractionated E. coli tRNA as a function of temperature. Conditions: 0.05 M sodium cacodylate, pH=6, 0.01 M MgCl$_2$. [(a) from Ref. 11; (b), (c), and (d) from Ref. 20.]

can be used to establish the sequence of the helical regions of unknown tRNA molecules.

Since certain regions of the tRNA molecules are accessible to chemical or enzymatic attack, it was possible to verify the genetic code by direct hybridization of the corresponding codon-triplets with the tRNA. Under well-defined conditions (high salt concentration, low temperature) only the correct codon-triplet will hybridize.

8.2.4 X-RAY STUDIES ON tRNA CRYSTALS

The low molecular weight and compact structure of tRNA encouraged crystallographers to try their luck. Indeed, tRNA does crystallize, but in numerous crystalline forms. Also, different tRNA's can cocrystallize or they can crystallize with 5 S RNA. The recent X-ray study by Kim et al. (22), which is certainly a major step forward, shows an L-shaped tRNA molecule [Fig. 8.10(a)]. This structure, however, does not account for many of the results from solution studies, like the protection of the -T-Ψ-C-G loop or of the -C-C-A terminus. Thus, the structure may not necessarily be that of native tRNA. It should not be forgotten that tRNA, in the absence of Mg, assumes an apparently stable form (20), which is biologically inactive. Also there is considerable evidence that aminoacyl-tRNA and peptidyl-tRNA have tertiary structures different from that of uncharged native tRNA.

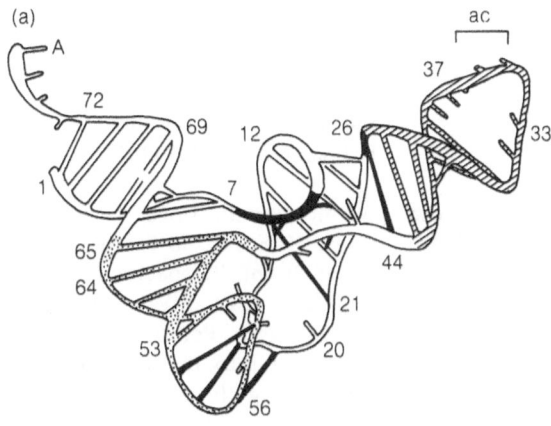

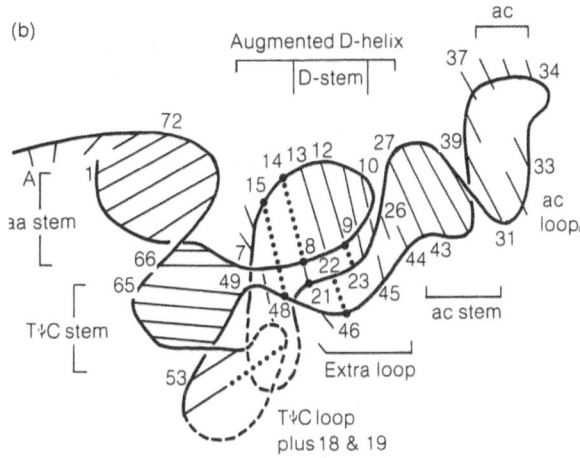

FIGURE 8.10. Diagrams of the crystallographically
determined structures of yeast tRNA$_{Phe}$. (a) Model of Kim
et al. (22); tertiary structure interactions (dark areas).
(b) Model of Robertus et al. (23); tertiary structure
interactions (dotted lines). In both models stacking of
bases and base pairs are indicated. The two models
have been placed in such a manner that the agreements
and differences between them can be visualized.

Figure 8.10(a) © 1973 by The American Association for the Advancement of
Science.

A more precise and detailed X-ray study of tRNA$_{phe}$ from
yeast has now been performed by Robertus et al. (23) Although the overall shape of the molecule resembles that
advanced by Kim et al. (22), many details are significantly
different [Fig. 8.10(b)]. Thus, the -T-Ψ-C-G loop seems to be
paired with the bases 18 and 19, the two Guo of the dihydro
Urd loop. The acceptor stem and the -T-Ψ-C-G helix form a
continuous helix, while the D-helix is perpendicular to

them. The anticodon stem is at a slight angle to the D-helix and is connected with it by a kind of hinge formed by the extra loop. Several additional base pairs and triples stabilize the structure, especially between the extra loop and the D-helix [see Fig. 8.10(b)].

8.3
Ribosomal RNA
and Ribosome
Structure

As mentioned above, 70 S ribosomes consist of two sub-units, 30 S and 50 S. The first contains a 16 S RNA (MW 600,000) and 21 specific proteins, whereas the second contains 23 S RNA (MW ~ 1.2 million), 5 S RNA, and 34 specific proteins. Removal of Mg causes the 70 S ribosome to dissociate into its 30 S and 50 S subunits (24). A major undertaking, the determination of large parts of the sequence of 16 S and 23 S RNA, was achieved (25). The 5 S RNA (MW 32,000), which is bound to the 50 S subunit, can only be obtained by unfolding the ribosome in highly concentrated LiCl solutions. In physiological salt solution, the 16 S and 23 S are single-stranded chains, with varying amounts of secondary structure due to back-folding of the chains upon themselves. In the ribosomal subunits, the RNA appears to be well protected by ribosomal proteins against nuclease attack (26,27).

The stability of the ribosomal complex depends critically on Mg concentration (26), which, in turn, depends on the physiological state of the ribosome. The subunits are firmly bound in translating ribosomes whereas free ribosomes are loosely associated, and a dynamic equilibrium exists among the 50 S, 30 S and 70 S particles. Organic divalent ions can replace Mg in ribosome stabilization, whereas alkali ions, particularly Li, have strong dissociating effects. Hydrophobic solvents (EtOH, DMSO, etc.) stabilize the ribosome, as do low temperature and low pH. EDTA and LiCl gradually deprive the ribosome of Mg, permitting fractionation of the ribosomal proteins. These proteins dissociate in groups, which can then be correlated with specific functions and locations in the ribosome (27). Another approach that localizes specific functions of protein synthesis on the ribosomal subunits involved the use of antibiotics and specific inhibitors (28).

Structural studies by biochemical, immunological, and physicochemical techniques are being actively pursued in many laboratories, directed towards establishment of a detailed three-dimensional map of the ribosome.

8.4
Viral RNA and
Virus Structure

Viruses, although very small, contain all the information needed for their reproduction. They consist of a protective shell of protein and a nucleic acid core, which may be RNA or DNA. The protein shell frequently contains enzymes capable of breaking the cell walls of living cells. The nucleic acid core enters the cell and directs the cell machinery to produce many copies of new virus particles, which are released by rupture of the cell (lysis) when a critical concentration has been reached.

The size of viruses varies between 100 and 2000 A. This is below the wavelength used in the light microscope, but viruses can be seen with the electron microscope or studied by X-ray diffraction. Data from the latter technique indicated quite early that the protein shell was symmetrically arranged around the nucleic acid core. This was the case for tobacco mosaic virus (TMV), in which the arrangement was found to be a helical rod (Fig. 8.11). Other viruses, which looked spherical in the electron microscope, were found to contain regular arrays of subunits with cubic sym-

FIGURE 8.11. Electron micrograph and model (insert) of TMV. The model shows the RNA core (dark dots) and the large bean-shaped protein subunits (all identical). The large particles represent full length viruses, which are infective.

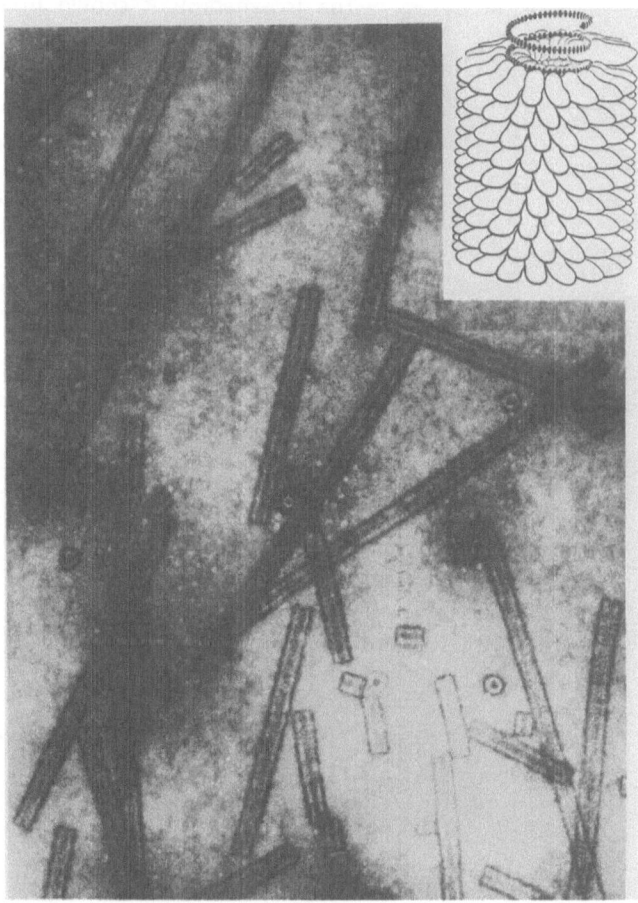

metry. This confirmed the suggestion by Crick and Watson (29) that the arrangement of (probably) identical subunits was best accomplished by cubic symmetry (see Fig. 8.11).

Electron microscopy contributed to the study of virus structure with two techniques: (a) heavy-metal shadowing, in which heavy metal vapors are deposited upon a grid with the virus particles; this reveals structural details where the metal atoms have been deposited. (b) Negative staining, in which the virus suspension is treated with phosphotungstate, an electron-dense material, which fills the empty spaces and crevices of the particle.

One distinguishes three classes of viruses: (a) helical (rod-shaped) like TMV (Fig. 8.11), (b) spherical (of cubic symmetry) like turnip yellow mosaic virus (TYMV), *Tipula iridescens* virus (Fig. 8.12), and herpes virus; and (c) complex symmetry like vaccinia or mumps virus and T-even phages (Fig. 8.13).

The spherical viruses are by far the largest group. Cubic symmetry includes the tetrahedron (4 triangles), the dodecahedron (12 pentagons), and the icosahedron (20 triangles). The icosahedron is shown in the example of the *Tipula irridescens* virus (an insect virus). Higher deltahedra are all derivatives of these three basic forms. [For a detailed dis-

FIGURE 8.12. Electron micrograph of *Tipula iridescens* virus (left) and shadow photograph of a regular icosahedron (b). The heavy metal shadowing at (a) and the shadow of the icosahedron are very similar and permit the conclusion that the virus is a regular icosahedron.

Figure 8.12 from Horne, "The structure of viruses."

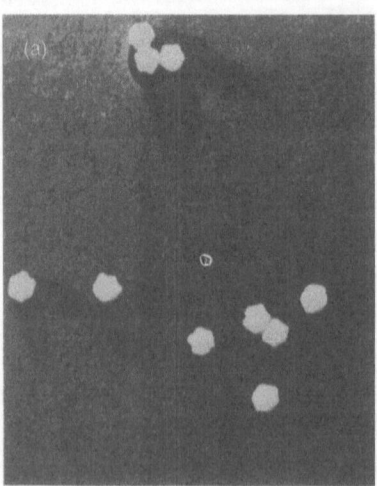
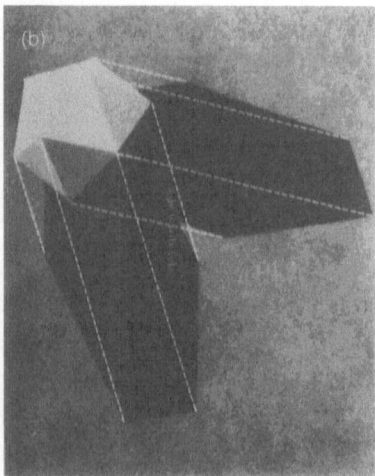

cussion see Caspar and Klug (30).] All the higher virus forms are assemblies of triangles arranged in five or six units on the vertex. A regular array of equilateral triangles arranged in hexamers will only be two-dimensional—hence the need for pentameric arrangements at regular distances.

FIGURE 8.13. Phage T$_2$ in its extended (a) and contracted (b) form. Note the elongated sheath in the extended form, which contains the DNA in the phage head; the head in the contracted form is empty.

Negative stained electron micrograph by E. Boy de la Tour. (From B. D. Davies *et al., Microbiology*, Harper & Row, 1967.)

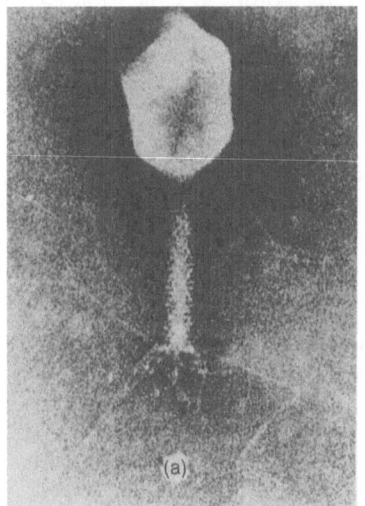
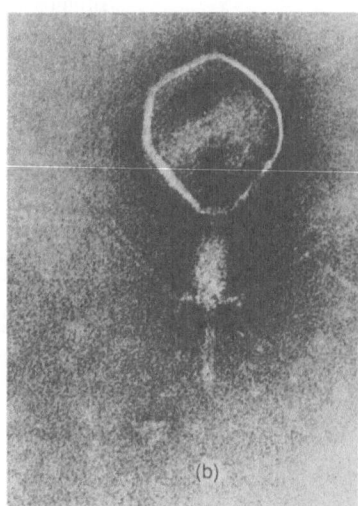

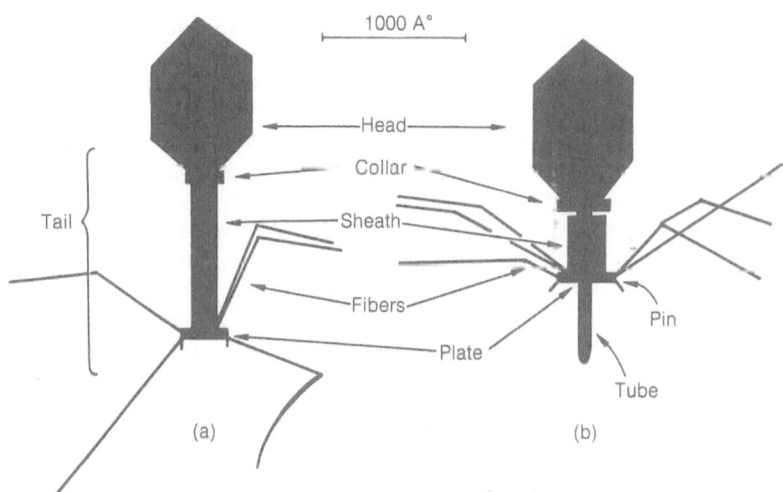

There are always 12 pentameric vertexes, plus a varying number of hexameric vertexes.

Little is known about the arrangement of the nucleic acid inside the virus, except in TMV, where it is a regular single-stranded helix with three bases for each of the 2130 identical capsomers (Fig. 8.11).

Details of the structure of DNA and RNA in various viruses are beginning to be unraveled. One of the most fascinating virus structures is that of the T-even phages. Figure 8.13 shows the complex symmetry of the T_2 phage. The tail, which is normally extended, can be contracted when the phage attacks the *E.coli* bacterium, thus injecting the T_2-DNA into the bacterial cell. The head of the T-even phages is an icosahedron, in which the central ring is doubled to form a hexagonal bipyramid.

References   1. Jacob, F. and Monod, J. (1961). *J. Mol. Biol.*, **3**, 318–56.

2. Brenner, S., Jacob, F., and Meselson, M. (1961). *Nature*, **190**, 576–81.

3. Gros, F. *et al.* (1961). *Nature*, **190**, 581–85.

4. Palade, G. E. (1955). *J. Biophys. Biochem. Cyt.* **1**, 59–68.

5. Crick, F. H. C. (1963). *Prog. Nucleic Acid Res. Mol. Biol.*, **1**, 163–217.

6. Keller, E. and Zamecnik, P. (1956). *J. Biol. Chem.*, **221**, 45–65; Holley, R. W. (1957). *J. Am. Chem. Soc.*, **79**, 658; Hoagland, M. B., Zamecnik, P. C., and Stephenson, M. L. (1957). *Biochim. Biophys. Acta*, **24**, 215.

7. Spirin, A. S. and Gavrilova, L. P. (1969). *The Ribosome.* Heidelberg, New York: Springer-Verlag.

8. Crick, F. H. C. (1966). *J. Mol. Biol.*, **19**, 548–58.

9. *Cold Spring Harb. Symp. Quant. Biol.*, **28** (1963); *ibid.*, **31** (1966).

10. Zachau, H. G. *et al.* (1966). *Cold Spring Harb. Symp. Quant. Biol.*, **31**, 417–24.

11. Fresco, J. R., Klotz, L. C., and Richards, E. G. (1963). *Cold Spring Harb. Symp. Quant. Biol.*, **28**, 83–90.

12. Holley, R. W. *et al.* (1965). *Science*, **147**, 1462–65; Holley, R. W. (1968). *Prog. Nucleic Acid Res. Mol. Biol.*, **8**, 37–48.

13. Dirheimer, G. *et al.* (1972). *Biochimie*, **54**, 127–44.

14. Cramer, F. (1971). *Prog. Nucleic Acid Res. Mol. Biol.*, **11**, 391–421.

15. Arnott, S. (1971). *Prog. Biophys.*, **22**, 181–213.

16. Hall, R. H. (1970). *Prog. Nucleic Acid Res. Mol. Biol.*, **10**, 57–85.

17. Favre, A., Yaniv, M., and Michelson, A. M. (1969). *Biochem. Biophys. Res. Comm.*, **37**, 266–70.

18. Thang, M. N. *et al.* (1967). *J. Mol. Biol.*, **26**, 403–21.

19. Chambers, R. W. (1971). *Prog. Nucleic Acid Res. Mol. Biol.*, **11**, 489–525.

20. Fresco, J. R. *et al.* (1966). *Cold Spring Harb. Symp. Quant. Biol.*, **31**, 527–37.

21. Kearns, D. R. *et al.* (1973). *J. Mol. Biol.*, **61**, 265–70; Wong, Y. P. *et al.* (1973). *J. Mol. Biol.*, **74**, 403–6.

22. Kim, S. H. *et al.* (1973). *Science*, **179**, 285–88; (1974). *ibid.*, **185**, 435–40.

23. Robertus, J. D. *et al.* (1974). *Nature*, **250**, 546–51.

24. Tissieres, A. *et al.* (1959). *J. Mol. Biol.*, **1**, 221–33.

25. Fellner, P., Ehresmann, C., and Ebel, J. P. (1972). *Biochimie*, **54**, 853–900; Ehresmann, C., Stiegler, P., Fellner, P., and Ebel, J. P. (1972). *Biochimie*, **54**, 901–68.

26. Spirin, A. S. (1974). *FEBS Letters Suppl.*, **40**, S38–47.

27. Pongs, O., Nierhaus, K. H., Erdmann, V. A., and Wittmann, H. G. (1974). *FEBS Letters Suppl.*, **40**, S28–37.

28. Vazquez, D. (1974). *FEBS Letters Suppl.*, **40**, S63–84.

29. Crick, F. H. C. and Watson, J. D. (1956). *Nature*, **177**, 473–75.

30. Caspar, D. L. D. and Klug, A. (1962). *Cold Spring Harb. Symp. Quant. Biol.*, **27**, 1–24.

9 Nucleic Acid-Protein Interactions

The field of nucleic acid–protein interactions has become increasingly active in the last few years, and it is thus difficult to be up-to-date here. This discussion will focus on the structural factors that determine the interactions between nucleic acids and proteins. [An excellent account of DNA–protein interactions has been written by von Hippel and McGhee (1).]

The entire mechanism of information transfer from DNA to protein sequences is controlled and made possible by interactions involving nucleic acids and proteins. This includes replication, transcription, translation, repressor mechanisms, and, presumably, differentiation.

We have to distinguish between two general types of interactions: specific and nonspecific. Specific interactions occur between a protein and a given nucleotide sequence on the nucleic acid molecule, i.e., where the binding constant for the specific site and for the rest of the molecule must differ by at least a factor of 10^4 to 10^6. Examples of this kind are polymerase–DNA interactions, the repressor–operator system, tRNA–synthetase complexes, nuclease binding, etc. Nonspecific interactions can be divided into two groups, those due to erroneous binding of the protein to other than specific sites, and those in which the interaction can take place anywhere with about the same efficiency; i.e., there is little or no difference in the binding constant for different sites on the nucleic acid. To this last category belong most of the polyamine and polypeptide interactions, as well as, to a lesser degree, histone–DNA complexes. Such interactions are readily understood in terms of simple physicochemical principles, such as electrostatic interactions between positively charged amino groups and negatively charged phosphates, combined with steric considerations.

The interaction between tRNA and its corresponding synthetase is, in this respect, much simpler, in particular due to the probably precise tertiary structure of tRNA and the relatively comparable, much smaller size of the two reactants. In this case the kinetic barrier is not prohibitive and a diffusion-controlled process could account for this reaction.

More complicated, and less obvious, is the mechanism for the precise recognition of base sequences in DNA by a given protein. Such recognition is an impressive feat kinetically, for an often unique nucleotide sequence (usually only a few base pairs long) must be located among tens or hundreds of thousands of similar sequences in DNA or RNA within a "reasonable" time (milliseconds!).

Diffusion-controlled processes cannot account for most specific DNA–protein interactions, in which the proper site cannot possibly be found by random collisions. A one-dimensional random walk along the DNA molecule is the most probable explanation. Here nonspecific binding to the DNA and local adsorption and desorption, with maintenance of many transient states through electrostatic interactions, is facilitated by the existence of grooves in the DNA helix, which considerably reduce the entropic barrier of the kinetics of the recognition process. Still, some processes are so rapid that special kinetic mechanisms have to be envisaged.

In any specific interaction, more than one attachment site must be involved, to allow for the specificity of the reaction and a sufficiently large free-energy change. For instance, in the case of the initiation binding of RNA polymerase to T_7 DNA, for each binding site (there are probably 3), about 20 base pairs out of 40,000 are recognized specifically. This implies that the binding constant for the specific site must be at least 2000 times larger than for the rest of DNA. This would correspond to a free energy difference between binding to specific and nonspecific sites of 4 to 5 kcal. Although this is a reasonably large value, it is not apparent that this free-energy difference would be sufficient to locally melt a region of 20 base pairs, as one might postulate for the action of RNA polymerase. In conjunction with the "breathing" of DNA (see Chapter 5, Section 5.2.4), these 4 to 5 kcal may be sufficient to open weaker, generally A-T rich regions.

Another problem is the relative frugality of possible recognition sites on the DNA helix. Apart from uniformly distributed negative phosphate charges, we can only consider the grooves as possible attachment sites. In the large groove we find for the A-T pair the N^7 and the amino group of Ade and the methyl group of Thy. In the case of the G-C pair, the amino group and C^5 of Cyt and the N^7 of Gua are the candidates. In the small groove the essential distinction between the two pairs is the amino group of Gua. It is possible and probable that the sequential accumulation of the same base pairs produces conformation changes in these regions (see Chapter 5, Sections 5.2.1–3 and Chapter 6, Section 6.2). But this is speculative and no precise molecular description exists. Still, such differences may be sufficient, or perhaps essential, to furnish some distinction between specific and nonspecific sites.

9.1 Polyamines

Spermine is the most extensively studied polyamine. It contains up to four positive charges per molecule and can bind tightly to DNA, competing for binding sites with magnesium ions. The optimal stoichiometry (at pH<8) is one spermine for four phosphates. It appears now, in contrast to previous reports, that there is no difference in the binding of spermine to DNA's of different base composition. The interaction frequently leads to precipitation when charge neutralization is approached. This fact has been used for the fractionation of tRNA's.

Recently, Gabbay and coworkers (2) have systematically investigated a great number of modified polyamines and their interaction with DNA. They have used certain of these compounds as a kind of molecular ruler and found, for instance, that the solution conformation of poly(A)·poly(U) is different from that of poly(I)·poly(C). Some of these "reporter" molecules can bind only in the minor groove. Since these compounds also bind to the same extent to chromatin (3), it appears that the histones in chromatin must be located in the major groove.

9.2 Polypeptides and Protamines

Many basic polypeptides (like polyLys or polyArg) bind to DNA. The large positive charge is responsible for the tight binding of these compounds, since high ionic strength is the best and simplest way to dissociate these complexes. Since very similar behavior is observed with DNA–histone complexes these polypeptides have been used as model systems to investigate unspecific DNA–protein interactions; 1:1 stoichiometry has been observed for these complexes as well as for double-stranded polynucleotide helices.

Several authors (4–7) have demonstrated that poly-Lys preferentially interacts with and even precipitates A–T–rich DNA. Polyarginine prefers G–C–rich DNA's. Tetramethylammonium ions reverse these bindings. The binding of these polypeptides appears to be cooperative, i.e., in subsaturating amounts of polypeptide, some DNA molecules are completely covered, while the rest remains "naked." Tritium-exchange experiments have shown that a conformational change in the DNA occurs and that about one-fourth of the slowly exchanging hydrogens (presumably involved in hydrogen bonds) become rapidly exchangeable.

That specific DNA–polypeptide complexes are formed is also evidenced by the characteristic ORD and CD spectra of these complexes (Fig. 9.1). Very similar spectra are obtained also with DNA–histone associations, an indication of the similarity of these interactions.

X-ray studies on nucleoprotamines (8) and DNA–polypeptide complexes (9) have indicated that DNA remains essentially in the B form at high relative humidities. By lowering the relative humidity these complexes do not undergo the usual transitions to the A or C form, but maintain the B form.

**9.3
Histones and
Chromatin**

Chromosomes are discrete structures, generally visible in the light microscope during meiosis or mitosis in eucaryotic cells. Chromatin is the more or less diffuse structure (from the cytological point of view) isolated from interphase nuclei. Isolated chromatin also is referred to as nucleohistone or nucleoprotein. It may contain material other than DNA and protein.

Whole bacterial chromosomes have been isolated by different groups (11,12) and were found to contain the entire DNA of the bacterial cell. Kavenoff and Zimm (12) recently obtained evidence that chromosomes from *Droso-*

FIGURE 9.1. CD spectra of poly-lysine-DNA complexes (a) and H1-histone-DNA complexes (b) at different polypeptide/DNA ratios *r* expressed in mole residues amino acid or nucleotide.

(a) 1: $r=0$, 2: $r=0.27$, 3: $r=0.54$, 4: $r=0.81$. (After Ref. 7.)

(b) 1: $r=0$, 2: $r=0.25$, 3: $r=0.5$, 4: $r=0.75$, 5: $r=1.0$. (After Ref. 6.)

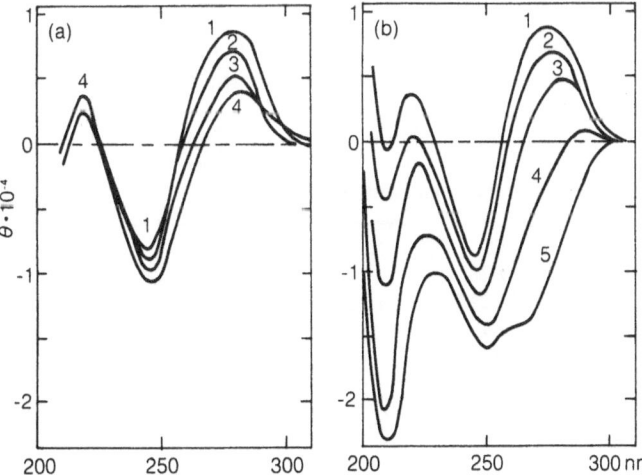

phila melanogaster probably contain only one or, at the most, two DNA molecules. Similar studies on yeasts have shown that the chromosomes are just as long as a single DNA molecule would be.

Histones are the basic proteins associated with the DNA of chromosomes of virtually all eucaryotic organisms. These proteins are remarkably similar in all organisms (amino acid sequence etc.). Five types of histones, classified according to their amino acid composition, are known (Table 9.1). Their sequences have all been determined, and they show that the basic amino acids are concentrated in one half of the molecule. The fact that they differ so little from one species to another suggests that they are indispensable for maintenance of eucaryotic chromosome structure. They are present in about equimolar amounts.

The H1 histone can selectively be removed from chromatin by 0.6 M NaCl; this produces distinct structural changes in the chromatin. It is possible that H1 histones are deposited on the surface of other histones. The attachment sites on the H1 histones appear to be the Lys-rich C and N termini; the central part may be free to interact with other H1 histones or other chromatin proteins. Neither H1, nor the H2A and H2B histones appear to be arranged in a tandem manner. The H3 is unique for its SH-groups which give rise to dimers and polymers in metaphase chromosomes. The Arg-rich histones, H3 and H4, are probably involved in maintaining the supercoiled structure of DNA in chromatin.

What role histones may have besides structural stabilization is far from clear. They may act as generalized gene suppressors, i.e., they may control the expression of the entire eucaryotic genome. Their removal increases the template activity of the DNA, for instance.

Another class of proteins bound to DNA are the non-

Table 9.1 Characteristics of Calf Thymus Histones and Different Nomenclatures Found in the Literature

Names	Nomenclature after Johns and Butler (13)	Single letter code[a]	New code (25)	Lys/Arg	M W
Lys-rich	F1	KAP	H1	20	20000
Slightly Lys-rich	F2B	KAS	H2B	2.5	13774
	F2A2	LAK	H2A	1.0–1.2	15000
Arg-rich	F3	ARE	H3	0.72	15324
Arg-rich	F2A1	GRK	H4	0.79	11282

[a] Nomenclature according to amino acid single letter code A: Alanine; E: Glutamic Acid; G: Glycine; K: Lysine; L: Leucine; P: Proline; R: Arginine; S: Serine.

histone proteins, frequently rich in glutamic acid. This class includes numerous enzyme activities, the precise role of which is unclear.

Chromosomes have been studied by a variety of methods, especially electromicroscopy, X-ray diffraction and scattering, optical methods, and hydrodynamic techniques. Chromosome fibers have been shown to undergo various structural changes (14–19) upon changes in relative humidity, but wide disagreement about their dimensions is found in the literature. Thus, fiber thickness determined by electron microscopy has been reported to be anywhere from 80 to 250 A. Similar disagreement exists with other methods. The critical factor in chromosome stabilization seems to be the presence of divalent ions like Mg or Ca (or alternatively very high ionic strength, >0.6 M NaCl).

The X-ray work of Pardon et al. (14–16) showed reflections at 105, 55, 35, 27, and 22 A; from these data they developed a model of a supercoil with about 130 A diameter and 120 A pitch. Bram and Ris (17,18) have deduced a much tighter supercoil with about 100 A diameter and 50 A pitch from X-ray scattering and electron microscopy data. Again, the presence of divalent ions seems to be critical.

Recent neutron scattering experiments (19) have clearly shown that the spacings at 105 and 37 A are due to the protein components of the chromatin, while those at 55 and 27 A are contributed by the DNA. This distinction has been made possible by "contrast matching" of either of the components (which have different proton densities and thus exhibit different neutron scattering) by varying the H_2O/D_2O ratio of the solvent. These results shed serious doubts on the supercoil models, but bring strong support to the string-of-beads structure observed by electron microscopy (20). These beads appear to contain the histones H2A, H2B, H3, and H4, surrounded by the DNA in a supercoil of 100 A diameter and 55 A pitch. The beads (v-bodies) are separated by several hundred A of apparently free DNA. It is proposed that Histone H1 is complexed with DNA on the outside of the spherical unit and may have a crosslinking role. These data are also in agreement with NMR and optical results (21) on histone association.

The ionic strength requirement for the maintenance of quaternary structure of chromosomic material has been frequently neglected in hydrodynamic and optical measurements. These techniques have been used to follow the reconstitution of chromosomes from DNA and the different histones. The differential melting of such partially dehistonized structures has been studied successfully, and the presence of Ca seems to be essential.

There is still disagreement about the need or requirement for certain histones to maintain the chromosome structure. This field is very active now, and within the next few years considerable progress is to be expected. The work done up to 1974 is reviewed in Refs. 22–25.

References

1. Von Hipple, P. and McGhee, J. D. (1972). *Ann. Rev. Biochem.,* **41**, 231–300.
2. Gabbay, B. J. (1969). *J. Am. Chem. Soc.,* **91**, 5136–50.
3. Simpson, R. T. (1970). *Biochemistry,* **9**, 4814.
4. Cohen, P. and Kidson, C. (1969). *J. Mol. Biol.,* **35**, 241–44.
5. Shapiro, J. T., Leng, M., and Felsenfeld, G. (1969). *Biochemistry,* **8**, 3219–26.
6. Fasman, G. D., Schaffhausen, B., Goldsmith, C., and Padler, A. (1970). *Biochemistry,* **9**, 2814–22.
7. Chang, C., Weiskoff, M., and Li, H. J. (1973). *Biochemistry,* **12**, 3028–32.
8. Subirana, J. A. and Puigjaner, L. C. (1973). *Jerusalem Symp.,* **5**, 645–652.
9. Suwalsky, M. and Traub, W. (1972). *Biopolymers,* **11**, 2223–31.
10. Stonington, O. G. and Pettijohn, D. E. (1971). *Proc. Nat'l. Acad. Sci., U.S.A.,* **68**, 6–9.
11. Worcel, A. and Burgi, E. (1972). *J. Mol. Biol.,* **71**, 127–47.
12. Kavenoff, R. and Zimm, B. H. (1973). *Chromosoma,* **41**, 1–27.
13. Johns, E. W. and Butler, J. A. V. (1962). *Biochem. J.,* **82**, 15–18.
14. Pardon, J. E., Wilkins, M. H. F., and Richards, B. M. (1967). *Nature,* **215**, 508.
15. Richards, B. M. and Pardon, J. E. (1970). *Exptl. Cell Res.,* **62**, 184–96.
16. Pooley, A. S., Pardon, J. F., and Richards, B. M. (1974). *J. Mol. Biol.,* **85**, 533–49.
17. Bram, S. and Ris, H. (1971). *J. Mol. Biol.,* **55**, 325–36.
18. Bram, S., (1972). *Biochimie,* **54**, 1005–11.
19. Baldwin, J. P. Boseley, P. G., Bradbury, E. M., and Ibel, K. (1975). *Nature* **253**, 245–49.
20. Olins, D. E. and Olins, A. L. (1972). *J. Cell Biol.,* **53**, 715–735; (1974). *Science,* **183**, 330–32.
21. Bradbury, E. M. et al. (1972). *Eur. J. Biochem.,* **26**, 482–489; (1972). *Eur. J. Biochem.,* **27**, 270–281; (1973). *Ann N.Y. Acad. Sci.,* **222**, 266–89.
22. (1974). *Cold Spring Harb. Symp. Quant. Biol.,* **39**.
23. Monahan, J. J. and Hall, R. H. (1974). *CRC Crit. Rev. Biochem.,* **2**, 67–112.
24. Huberman, J. A. (1973). *Ann. Rev. Biochem.,* **42**, 355–378.
25. (1975). Structure and function of chromatin, *CIBA Foundation Symposia #21,* London.

In the Guise of
an Epilogue

The subject of nucleic acid–protein interactions is one of the most fascinating areas of molecular biology today, and it is a good topic on which to close this book on nucleic acid structure.

In any kind of human endeavor, it is generally more difficult to finish than to start. Finishing a book is an arbitrary process, but science will continue as long as human curiosity exists.

Mozart could have continued his A Major violin concerto. The solo violin takes up a previous theme, repeats it twice, and could go on. Mozart chose to finish abruptly with a soft upward movement of the soloist. But music (like science) continues. . . .

Appendix:
Abbreviations and
Symbols for
Nucleic Acids,
Polynucleotides,
and Their
Constituents

IUPAC-IUB Commission on
Biochemical Nomenclature (CBN),
Recommendations 1970[1]

INTRODUCTION

The 1965 Revision of *Abbreviations and Symbols for Chemical Names of Special Interest in Biological Chemistry* was completed and published in 1965 and 1966 [1], almost coincident with the elucidation of the first complete nucleic acid sequence [2,3] and with the development of methods for the synthesis of specific polynucleotide sequences [4]. The latter developments and others (*e.g.*, modification of sugar components, synthesis of unnatural linkages) require a unified system for representing long sequences containing unusual or modified nucleoside residues. The system should facilitate comparisons between two or more such extended molecules, as in the search for homologies. At the same time, it must retain sufficient flexibility to accommodate the large variety of polymers synthesized by polymerases and be consistent, in this regard, with the rules governing the representation of polymerized amino acids [5].

The workers who first encountered these various needs invented a number of devices to achieve the representations required in their own papers, basing these for the most part upon the one-letter system presented in Section 5.4 of *Abbreviations and Symbols* [1]. Few of these devices have the capability of meeting all the situations that are now apparent. Hence the effort was undertaken to construct a system meeting as many of the latter as possible, preserving the previous, basic system and introducing additional conventions. This effort, as did the previous one, involved consultation with a large number of active workers in many countries over a period of some years. The conventions added here (indicated by ▲ for major additions, △ for minor revisions) are already in use by many of them, *e.g.* [3,6—8].

The present (1970) Recommendations are the result; they replace Section 5 of the previous Tentative Rules [1].

N-1. Abbreviations

N-1.1. SIMPLE NUCLEOTIDES[2]

The 5'-mono-, di-, and triphosphates of the common ribonucleosides may be represented by the customary abbreviations exemplified by AMP, ADP, ATP (AtetraP) in the adenosine series. The corresponding derivatives of other nucleosides are abbreviated similarly, using the

[1] These Recommendations were approved by the IUPAC-IUB Commission on Biochemical Nomenclature in 1969 and are published by permission of IUPAC and IUB.

Comments and suggestions for future revisions may be sent to any member of CBN: O. Hoffmann-Ostenhof (Chairman), A. E. Braunstein, W. E. Cohn (Secretary), P. Karlson, B. Keil, W. Klyne, C. Liébecq, E. C. Slater, E. C. Webb, and W. J. Whelan.

Reprints of these Recommendations may be obtained from the NAS-NRC Office of Biochemical Nomenclature (Dr. Waldo E. Cohn, Director), Biology Division, Oak Ridge National Laboratory, Box Y, Oak Ridge, Tennessee 37830, U.S.A.

△ symbols in N-3.2, *i.e.*, A, C, G, I, T, U, Ψ, X for the known nucleosides; R and Y for unspecified purine and pyrimidine nucleosides, respectively; N for unspecified nucleoside (no X or Y); B, S, and D are reserved for 5-bromouridine, thio uridine, and 5,6-dihydrouridine, respectively. Orotidine may
△ be designated by O to give OMP for orotidine 5'-phosphate[1]
△ The di- and triphosphates may on occasion be better expressed in the alternate form ppN or pppN, as in the poly merization equation n ppN → (pN)$_n$ + n P$_i$, or when the outcome of specific labeling is to be indicated, *e.g.*, n ppp$\overset{*}{N}$

→ (pN)$_n$ + n PP.

Uridine diphosphate glucose may be represented as UDPG or UDP-Glc; the latter form is preferred if there is the possibility of confusing G for glucose with G for guanosine.

In the context of the chemistry of the nucleosides or nucleotides, the more systematic three-letter symbols (N-2) should be used, *e.g.*, Ado-5'PPP or Urd-5'PP-Glc (N-2.4.3)

N-1.2. NUCLEOTIDE COENZYMES AND RELATED SUBSTANCES

Riboflavin 5'-phosphate (flavin mono-nucleotide)	FMN
Flavin-adenine dinucleotide (oxidized and reduced)	FAD, FADH$_2$
Nicotinamide mononucleotide	NMN
Nicotinamide-adenine dinucleotide[3] (oxidized and reduced)	NAD+, NADH
Nicotinamide-adenine dinucleotide phosphate[4]	NADP+, NADPH

△ Analogues of NAD or NADP (the generic terms requiring neither the plus sign nor the H) may be named by sub stituting an appropriate defined symbol for the N or the A

▲ [2] When abbreviations for single bases or nucleosides are required and permitted, the three-letter symbols listed in N-2.2 and N-2.3 should be used (see Comments in those sections), and not single letters and not, *e.g.*, UR, TdR, etc. Examples:

	Proscribed	Proposed
Fluorouracil	FU	FUra
Fluorouridine	FUR	FUrd
Fluorodeoxyuridine	FUdR	FdUrd
Thymidine	TdR	dThd
Bromouracil	BU	BrUra
Bromodeoxyuridine	BUdR	BrdUrd

[3] Formerly diphosphopyridine nucleotide (DPN, DPN+, DPNH) and coenzyme I.
[4] Formerly triphosphopyridine nucleotide (TPN, TPN+, TPNH) and coenzyme II.

e.g., AcPd (for acetyl-pyridine) in place of N; I (for inosine) in place of A, etc.

Semi-systematic names (see N-2) may often be used to advantage in discussing the chemistry of these dinucleotides, *e.g.*, NADP = Nir-5′-PP5′-Ado-2′P.

N-1.3. NUCLEIC ACIDS

N-1.3.1. The two main types of nucleic acids are designated by their customary abbreviations, RNA (ribonucleic acid or ribonucleate) and DNA (deoxyribonucleic acid or deoxyribonucleate). Ribonucleoprotein and deoxyribonucleoprotein should not be abbreviated.

N-1.3.2. *RNA Fractions*

Fractions of RNA or DNA, or functions exercised by preparations of RNA may be designated as follows:

messenger RNA	mRNA	transfer RNA[5]	tRNA
ribosomal RNA	rRNA	complementary RNA	cRNA △
nuclear RNA	nRNA	mitochondrial DNA	mtDNA △

These are generic terms and apply to preparations as well as to specific molecules.

N-1.3.3. *Transfer RNA's*

Transfer RNA's that accept a specific amino acid are designated as follows (using alanine tRNA as an example):
a) Nonacylated: alanine tRNA or tRNA^{Ala};
b) Aminoacylated: alanyl-tRNA or Ala-tRNA, or Ala-tRNA^{Ala}.

Comment. (i) The hyphen (in b) represents the aminoacyl bond and should not be used to connect a noun-adjective; (ii) the attached aminoacyl residue (in b) has the -yl ending, whereas the adjective describing the nonacylated form (a) does not; (iii) the superscript designator utilizes the conventional symbols for amino acid residues [1,9] exactly — one capital, two small letters.

Isoacceptors, *i.e.*, two or more tRNA's accepting the same amino acid, are designated by subscripts, *e.g.*, tRNA$_1^{Ala}$, tRNA$_2^{Ala}$, etc.

Specification of source may be made in parentheses before or after the abbreviation, *e.g.*, (*E. coli*) tRNA^{Ala}, alanyl-tRNA^{Ala} (*E. coli*).

△   The special problem of the particular methionine tRNA (tRNA^{Met}) that, once aminoacylated to give Met-tRNA, can be formylated to fMet-tRNA may be solved by the use of a subscript f (in the isoacceptor position) or by the use of tRNA^{fMet}. Thus tRNA$_f^{Met}$ (or tRNA^{fMet}) can be converted enzymically to Met-tRNA$_f^{Met}$ (or Met-tRNA^{fMet}) and then to fMet-tRNA$_f^{Met}$ (or fMet-tRNA^{fMet}); Met-tRNA^{Met} cannot be formylated enzymically.

Symbols

General Concepts and Conventions

Two systems are recognized, designated the "three-letter" and the "one-letter" system, respectively. The first (N-2), patterned after the systems in use for amino acid and saccharide residues in polymers [1], is designed largely for descriptions of chemical work involving bases, nucleosides, nucleotides and very small oligonucleotides, or for abbreviating these in minimum space (as on chromatograms or figures or table headings). The "one-letter" system (N-3 and N-4) is designed for the representation of oligonucleotides or polynucleotides, or parts thereof, and for their noncovalent associations, not for mononucleotides or nucleotides. Neither system is intended to replace the names of the latter substances in the text of papers.

In both systems, it is assumed, in the absence of appropriate symbols, that (a) all nucleosides (except pseudouridine) are 1-(pyrimidine) or 9-(purine) glycosyls, (b) all nucleoside linkages are β, (c) all sugar configurations are D, (d) all sugar residues are ribosyls unless otherwise specified, (e) all deoxyribosyls are 2′-deoxyribosyls, and (f) only 3′ → 5′ linkages, read from left to right, are involved.

---

[5] Replaces "soluble" RNA (sRNA), which should no longer be used for this purpose. RNA soluble in molar salt, or nonsedimentable at $100000 \times g$, or exhibiting a sedimentation coefficient of 4 S, should not be termed sRNA.

N-2. THREE-LETTER SYMBOLS[6]

N-2.1. *Phosphoric Acid Radical*

The phosphoric acid radical, whether monoesterified or diesterified, is designated by an italic capital P.

△   N-2.2. *Purines and Pyrimidines*

These are designated by the first three letters of their trivial names:

Ade	adenine	Thy	thymine
Gua	guanine	Cyt	cytosine
Xan	xanthine	Ura	uracil
Hyp	hypoxanthine	Oro	orotate
Pur	unknown purine	Pyr	unknown pyrimidine
	Base	unknown base	

Sur and Shy may be considered for thiouracil and thiohypoxanthine (6-mercaptopurine), respectively.

When abbreviations for single purines or pyrimidines are required and permitted, the above symbols should be used rather than A, C, G, T, U, etc.[2].

N-2.3. *Nucleosides*

N-2.3.1. *The ribonucleosides* are designated by the following symbols, chosen to avoid confusion with the corresponding bases:

Ado	adenosine		Thd	ribosylthymine (not
Guo	guanosine			thymidine)
Ino	inosine		Cyd	cytidine
△ Sno	thioinosine (mercapto-		Urd	uridine
	purine ribonucleoside)		Srd	thiouridine △
△ Xao	xanthosine		Ψrd	pseudouridine
△ Puo	"a purine nucleoside"		Ord	orotidine △
△ Nuc	"a nucleoside"		Pyd	"a pyrimidine nucleo- △
				side"

Ribosylnicotinamide may be designated by Nir.

Comment. The prefix r (for ribo) may be used for emphasis or clarity. It may precede a single residue or, if applicable, a connected series.

N-2.3.2. *The 2′-deoxyribonucleosides* are designated by the above symbols (N-2.3.1) prefixed by d, *e.g.*, dAdo for 2′-deoxyribosyladenine (deoxyadenosine), dThd for 2′-deoxyribosylthymine (thymidine). The d may be used as a prefix to a connected series if all members of that series are 2′-deoxyribosyl derivatives. In mixed series, r and d should both be used before the appropriate residues, *e.g.*, P-dAdo-P-rThd-P.

△   Other sugar residues may be indicated by similar prefixes, *e.g.*, a for arabinose, x for xylose, l for lyxose.

Comment. For special purposes, the base and the sugar may be designated separately, using the base abbreviations of N-2.2 and the standard sugar abbreviations [1], *i.e.*, Rib, Ara, Glc, etc. Thus, adenosine = Ado = Ade-Rib; thymidine = dThd = Thy-dRib. (The "de" used in section 3.5 of *Abbreviations and Symbols* [1] for deoxy may be shortened to "d" in this context).

When abbreviations for single nucleosides are required and permitted, the above symbols should be used, *e.g.*, Urd (not UR, Ur or U) and dThd (not TdR, Tdr, TDR, T or dT), for uridine and thymidine, respectively[3].

N-2.4. *Nucleotides*

N-2.4.1. *Mononucleotides*. In the three-letter symbols, mononucleotides are normally expressed as phosphoric esters, such as Ado-3′-P or P-3′-Ado for adenosine 3′-phosphate, P-2′-Guo or Guo-2′-P for guanosine 2′-phosphate, Cyd-5′-P or P-5′-Cyd for cytidine 5′-phosphate (see N-2.4.4).

N-2.4.2. *Cyclic phosphodiesters* are designated by two primed numerals, one for each point of attachment, as in Cyd-2′:3′-P (or P-2′:3′-Cyd) or in Ado-3′:5′-P (or P-3′:5′-Ado). (The corresponding bisphosphates would be Cyd-2′,3′-P_2 and Ado-3′,5′-P_2.)

△   N-2.4.3. *Nucleoside diphosphate sugars*, which center about a pyrophosphate group, are represented by, *e.g.*, Urd-5′PP-Glc for uridine diphosphate glucose, *i.e.*, uridine

---

[6] The IUPAC Commission on Nomenclature in Organic Chemistry prefers these symbols to the one-letter ones (N-3), designed for polymer representation. The three-letter symbols should be used whenever chemical changes involving nucleosides or nucleotides are being discussed.

5'-(α-D-glucopyranosyl diphosphate), often termed UDPG or UDP-Glc (see N-2.4.4 and N-1.1).

△   N-2.4.4. *Points of attachment* in oligo- or polynucleotides are designated by primed numerals, *e.g.*, 2'P5', 5'P5', etc. as in Ado-2'P5'-rThd-2'P or Ado-5'PP5'-Nir (for NAD; see N-2.4.3 and N-1.2). The positional numerals may precede a series, as in (2'-5')Ado-*P*-Guo-*P*-Urd-*P* to specify Ado-2'P5'-Guo-2'P5'-Urd-2'P. They may be omitted when the series in the left-to-right direction is 3'P5'.

Comment. Phosphate groups at the ends of chains may appear without numerals. In this case it is understood that *P*- at the left end means a 5'-phosphate, -*P* at the right means a free 3'-phosphate. Thus AMP can be represented by Ado-5'-*P*, *P*-5'-Ado, or *P*-Ado, but not by Ado-*P* (which would represent the 3'-phosphate).

N-3. ONE-LETTER SYMBOLS

N-3.1. PHOSPHORIC ACID RESIDUES

A monosubstituted (terminal) phosphoric residue is represented by a small p. A phosphoric diester (internal) in 3'-5' linkage is represented by a *hyphen* when the sequence is *known*, or by a *comma* when the sequence is *unknown*. Unknown sequences adjacent to known sequences are placed in parentheses; these replace, at the points where they occur, the need for other punctuation. All these symbols thus replace the classical 3'-5' or 3'p5' symbols (*cf.* N-3.3.1 and N-3.3.2). A 2':3'-cyclic phosphate residue may be indicated by > or >p.

Comments

i) The terminal p's should be specified unless their presence is unknown, in doubt, or of no significance to the argument.

ii) "Polarity" (direction other than 3'→ 5') is dealt with in N-3.3.2.

iii) Linkages other than 3' and 5' are specified by other means (see N-3.3.1).

iv) A codon triplet, in which definite left-to-right order and 3'-5' linkages are assumed and in which the termini are not of importance, may be written without punctuation, *e.g.*, AGC.

N-3.2. NUCLEOSIDES

N-3.2.1. *Ribonucleosides*[2]

The *common ribonucleoside residues* (radicals) are designated by single capital letters, as follows:

A adenosine	T ribosylthymine (not thymidine)
G guanosine	
I inosine	C cytidine
X xanthosine	U uridine
	Ψ pseudouridine[7]
△ R unspecified purine nucleoside	Y unspecified pyrimidine nucleoside

N unspecified or unknown nucleoside (do not use X, P, or any of the above).

Rare Nucleosides. It is often advantageous, *e.g.*, in comparing long sequences, to represent every nucleoside residue by a single letter rather than by a group of letters and numbers. In such cases, those capital letters not assigned to common nucleosides (above) may be arbitrarily defined and used. It is recommended that the following be reserved for the substances listed (*cf.* N-4.4):

△ D 5,6-dihydrouridine	B 5-bromouridine
△ S thiouridine (for locants, see N-4.4)	O orotidine (see N-1.1)

Other symbols for these and for other modifications are listed in N-4.

Comments

i) The prefix r for ribo should be used when there is need for the additional specification.

ii) Other sugars or modified sugars are considered in N-3.2.2, N-3.2.3 and N-4.2.

N-3.2.2. *Deoxyribonucleosides*

The common 2'-deoxyribonucleosides are designated by the above symbols, modified in one of the following ways:

---
[7] Q may replace Ψ for computer work.

a) When space is available and no other prefixes are required, the *prefix* d is used; thus (i) dA-dG-dC . . . or d(A-G-C . . .); (ii) poly[d(G-C)] or poly(dG-dC) (these are identical substances); d may precede each residue or a whole chain, as applicable.

▲   b) When space is available but other, possibly confusing, prefixes are involved, a *subscript* d is used; thus, mmtT$_d$-bzA$_d$-T$_d$-anC$_d$ for a protected tetradeoxynucleotide [4]. (The prefixes are defined in N-4.1.)

▲   N-3.2.3. *Unusual Sugar Residues*

Sugar moieties other than ribosyl or 2'-deoxyribosyl may be indicated as described in N-3.2.2 above, depending on requirements for base-modifying prefixes (N-4.1) and space available, using a, x and l (see N-2.3.2) for the other pentosyls, *ad hoc* letters for others, each defined; thus -aC- or -C$_a$- for an arabinosylcytosine residue. Symbols for substituents on sugars are given in N-4.2 (see also N-4.4).

N-3.3. OLIGO- AND POLYNUCLEOTIDES

N-3.3.1. *Points of Attachment*

The diesterified phosphate residue, represented by hyphen or comma (*cf.* N-3.1) is considered to be attached to the oxygen atom of the 3' carbon on its left and to that of the 5' carbon on its right. For other types of linkage, the simple hyphen must be replaced by its numerical form, as in 2'-5' (or 2'p5'), 5'-5', etc. [6], *e.g.*, G3'p5'A2'p5'A or G3'-5'A2'-5'A. These locants must precede a chain or a polymer if the internucleotide linkage is identical throughout, *e.g.*, (2'-5')A-U-G-C for the corresponding tetranucleotide.

▲   N-3.3.2. *Direction of the Phosphodiester Link*

The hyphen used in known sequences is a contraction of the arrow (→) that is understood to point to the 5' terminus of the phosphodiester bond (unless other numerals are used, as in N-3.3.1). When left-to-right direction is *not* the case, this must be indicated by an appropriate locant preceding the chain, or by an arrow to indicate the 3'→ 5' direction, as in the peptide rules [9]. Thus, associated hydrogen bonded segments (see N-3.4.2) may be represented by, *e.g.*,

$$(3'\text{-}5')\text{A-C-A-C-A-C etc.}$$
$$(5'\text{-}3')\text{U-G-U-G-U-G etc.}$$

or by

$$\text{A}\to\text{C}\to\text{A}\to\text{C}\to\text{A}\to\text{C etc.}$$
$$\text{U}\leftarrow\text{G}\leftarrow\text{U}\leftarrow\text{G}\leftarrow\text{U}\leftarrow\text{G etc.}$$

Another device used to represent "reverse polarity" is rotation of the symbols [10,11]. Thus the above associated polymers may be shown as

$$\text{A-C-A-C-A-C etc.}$$
$$\text{Ո-Ց-Ո-Ց-Ո-Ց etc.}$$

In such representation, the left-to-right 3'-5' convention is assumed to hold when the letters appear right-side up.

Examples of Oligonucleotides

A-G-Up (for ApGpUp); 3'→ 5' trinucleotide, terminal 3' phosphate.

A-G-U>p; the same, with terminal 2':3'-cyclic phosphate.

pA-G-U; the same, commencing with a 5' phosphate, terminating in an uridine with unsubstituted 2' and 3' hydroxyls.

△   pppG-G . . . Ap; this nucleotide (of unspecified length and sequence) has a 5'-triphosphate residue on the G at one (the 5') end and a 3'-phosphate on the A at the other (the 3') end.

pG-A-Ψ(C$_3$U)T-C-C-A; a decanucleotide, commencing (5' end) with a 5' phosphate, including a trinucleotide of unknown sequence between the Ψ and the T, and terminating (3' end) in an adenosine residue with unsubstituted 2' and 3' hydroxyl groups.

d(pG-A-C-T); tetranucleotide (all deoxy), with 5' terminal phosphate on G.

d(T̄←C̄←Ā←Gp); the same (arrow indicates 5'←3' direction).

pG-A$_d$-C-T; the same, but with two deoxy, two ribo residues (see N-3.2.2 b).

(2'-5')pG-A-C-T; the same, all ribo, all in 2'-5' linkage.

pG2'-5'A-C-T; the same, with a single 2'-5' linkage (between G and A).

AGC; a codon (Note: The symbols for phosphoric acid residues may be omitted in describing codons. This is an exception to N-3.1).

N-3.4. POLYMERIZED NUCLEOTIDES

N-3.4.1. *Single Chains*

Polynucleotides composed of repeating sequences or of unknown sequence may be represented by either of two systems essentially identical with those devised and recommended by the IUPAC Commission on Nomenclature of Macromolecules and by the American Chemical Society's Polymer Nomenclature Commission (see also *Synthetic Polypeptides* [5]).

a) The repeating unit is preceded by "poly", meaning "polymer of". Thus, polynucleotide or poly(N); polyadenylate or poly(A); poly(adenylate-cytidylate) or poly(A-C) (alternating); or poly(adenylate, cytidylate) or poly(A,C) (random).

b) The repeating unit, enclosed in parentheses if complex, is followed by a subscript denoting length, *e.g.* a number $(A-C)_{50}$, an average number $(A-C)_{\overline{50}}$ or a range $(A-C)_{40-80}$, if desired. Where the number of residues has not been determined and this form is required by the context, the subscript "n" may be used (as in ref. [5]). However, two n's should not appear in the same formula unless equal length is implied. When equal length is not the case, additional letters should be used, such as m, k, j, etc.

In either case, the symbols may carry prefixes or subscripts as required for proper specification. Note that "poly" is not used in the second system.

Examples

poly(A-U), alternating copolymer of A and U [12]; poly(A,U), random copolymer of A and U; *not* poly AU or poly $A + U$;
poly(A_2,U), as above but 2:1 in average composition; $(A_2,U)_{\overline{50}}$, as above, average length of chain, 150 residues; poly[d(A-T)] or poly(dA-dT), for alternating dA and dT[a] (see N-3.1 and [12]).

Comment. Multiple parentheses or brackets may be used for blocks within polymers, and vertical lines for side chains, etc. [5,9]. "Oligo" may replace "poly" where applicable. Terminal phosphate residues need not be specified unless they are essential to the argument.

N-3.4.2. *Association between Chains*

Association (noncovalent) between two or more polynucleotide chains, such as that ascribed to hydrogen-bonding, is indicated by the *center dot* (not the hyphen, which indicates covalent linkage), *e.g.* (*cf.* [12,14]):

a) poly(A) · poly(U), not poly(A · U), nor poly AU, nor poly $A + U$[b]; poly(A · U) may be used when it is implied that each A is paired with a U, regardless of chain lengths.

b) poly(A) · 2 poly(U) not poly(A · 2U), nor poly(A · U₂); poly(A · 2U) indicates the same triple-stranded complex and that each A is matched by two U's, regardless of individual chain lengths.

c) poly[d(A-T)] · poly[d(A-T)] or poly[d(A-T) · d(A-T)].

d) A · poly(U) or A · (U)ₙ for single adenosine residues associated with polyuridylate or poly(uridylic acid).

Absence of association between chains is indicated by the *plus* sign (traditional in chemistry for coexisting but nonassociated species) *e.g.*:

a) poly(dC) + poly(dT), not poly(dC + dT);
b) poly(dA,rT₄) + poly(dG);
c) 2 [(poly(A) · poly(U)] ⇌ poly(A) · 2 poly(U) + poly(A) [12].

The *absence of definite information on association* is indicated by the *comma* (as before, indicating "unknown"), *e.g.*:

a) poly(A), poly(A,U);
b) poly[d(G-C)], poly[d(A,T)].

---

[a] Poly[d(A-T)] or poly(dA-dT) was originally [13] termed poly dAT. While this has the advantage of brevity, it has proven ambiguous (see footnote [b]) in other situations and is inconsistent with the general principles of polymer symbolism (*e.g.* [5]). Hence, its use is not recommended.

[b] Poly AU and AU, etc., have been used for poly(A) · poly(U) [15]. The similarity of this system for associated homopolymers to that originally proposed for alternating copolymers (see foot-note [a]) can lead to confusion, in that it indicates one covalent chain rather than two. Its use is not recommended. Similar potential confusion attends the use of the other incorrect terms given in N-3.4.2.

Comments

i) Hyphens are *not* used for association (noncovalent); poly(A-U) specifies a single chain, not two chains.

▲ ii) The *center dot* should always be used to indicate base pairs involved in noncovalent associations (see N-3.3.2), *e.g.*, A · T base pair, or G · C hydrogen bonds (not A-T, or G-C which indicate covalent linkages). The center dot is located as shown, *above* the line.

▲ iii) In describing *base ratios*, the form $(A + T)/(G + C)$ should be used, not AT/GC, nor $A + T/G + C$. Two capital letters should not be juxtaposed (except as in N-3.1, comment iv), to distinguish sequence G-C, from content $G + C$, from ratio G:C or G/C, from base pair G · C.

▲

N-4. MODIFIED BASES, SUGARS, OR PHOSPHATES IN POLYNUCLEOTIDES

N-4.1. *Designation of Substituents on Bases*

In long sequences, as in transfer RNA's, where it is preferable to have not more than one capital letter per nucleoside residue, the standard symbols for nucleosides [*i.e.*, A, U, G, C, etc. (see N-3.2.1)] may be modified by a symbol of lower case letter(s) placed immediately before the single capital letter. Those symbols recommended for more common modifications are listed below (for locants and multipliers, see N-4.4; for unusual sugar residues, see N-3.2.2 and N-3.2.3:

m, e, ac	methyl, ethyl, acetyl
n, o	amino (N replaces H), deamino (O replaces N)
z, c	aza (N replaces C), deaza (C replaces N)
h	dihydro (hU = dihydrouridine; see also N-3.2.1 and N-4.4)
hm, ho (or oh)	hydroxymethyl, hydroxy
aa	aminoacyl
f	formyl (as in the conventional fMet for formylmethionyl)
fa	formylaminoacyl
i	isopentenyl (= γ,γ-dimethylallyl)
s	thio or mercapto (sU = thiouridine; see N-3.2.1 and N-4.4)
fl, cl, br, io	fluoro, chloro, bromo, iodo (not encountered in natural polynucleotides; see also N-3.2.1 and N-4.4).

Symbols for some N-protecting radicals used in synthetic work [4,9] are:

bz, bzl, tos	benzoyl, benzyl, tosyl
tr, an, bh	trityl, anisoyl, benzhydryl (diphenylmethyl)
mmt	monomethoxytrityl (*p*-anisyldiphenylmethyl)
dmt	dimethoxytrityl (di-*p*-anisylphenylmethyl)
thp, dns	tetrahydropyranyl, dansyl
cmc	N-cyclohexyl-N′-[β-(4-methylmorpholino) amidino] (reaction product from the corresponding carbodiimide [16].

In simpler situations where the avoidance of multiple capital letters in a single residue symbol seems not to be necessary, the standard chemical symbols (Me, Br, etc.) may be used. In such cases, no punctuation should appear between modifier and nucleoside symbol, *e.g.*, 6Me₂A, 5BrU. The prefix "di" should not be used; subscripts numerals suffice (*cf.* N-4.4).

Comments

i) Symbols for other protecting groups may be constructed according to the principles indicated here and in Section 6 of *Amino Acids and Peptides* [9].

ii) When space is severely restricted, these symbols may appear above the nucleoside symbol (see N-4.4) [3,4,7,8], *e.g.*,

$$\overset{\text{ac}}{\text{C}} \quad \text{for} \quad \text{acC.}$$

iii) Symbols for bifunctional adducts must lie above or below the chain (or chains) (see conventions for branched peptides in [5] and [9]) and hence may utilize any appropriate symbols. Thus a methylene bridge between two adenosines [17] could be represented as

$$\begin{matrix} -A- \\ | \\ CH_2 \\ | \\ -A- \end{matrix} \quad \text{or} \quad \begin{matrix} & CH_2 \\ & | \\ -A \cdots A- \end{matrix}$$

for inter- or intra-chain linkage, respectively.

▲   N-4.2. *Designation of Substituents on Sugars*

N-4.2.1. *Internal Modifications.* The symbols are *lower case* when the modified sugar is internal; they are placed immediately *to the right* of the nucleoside symbol and indicate substitution at the (internal) 2′ position unless otherwise specified. Thus -Am- indicates a 2′-*O*-methyl-adenosine residue [7,8] (see also N-4.4).

N-4.2.2. *Terminal Radicals.* The common, natural termini, phosphate and hydroxyl, are represented, if necessary, by p (N-3.1) and oh or ho (N-4.1); the latter is only required for emphasis as it is implied in the nucleotide symbol itself.

Other terminal radicals (hydroxyl-substituents) may utilize standard chemical symbols or abbreviations. These are placed in parentheses (following the appropriate nucleoside symbol, as noted above). Recommended abbreviations (aside from normal chemical symbols) are [4,9]:

(EtOEt), (EtOMe)	1-ethoxyethyl, ethoxymethyl
(Ph$_2$CH), (Bzl), (Tr)	benzhydryl, benzyl, trityl
(MeOTr), [(MeO)$_2$Tr]	monomethoxytrityl, dimethoxytrityl
(Me), (Et), (Ac), (Tos)	methyl, ethyl, acetyl, tosyl
(Thp), F$_3$CCO-	tetrahydropyranyl, trifluoroacetyl
(AA), (Gly), (Leu), etc.	aminoacyl, glycyl, leucyl, etc.

Terminal glycol-protecting (bifunctional) radicals, bridging the 2′ and 3′ hydroxyls unless otherwise indicated, may require the following:

(>CMe$_2$)	isopropylidene; *e.g.*, -C-C-A(>CMe$_2$)
(>BOH), (>CO)	borate, carbonyl
>p or >	2′:3′-phosphate (cyclic) (*cf.* N-3.1)

▲   N-4.3. *Phosphoric Acid Protecting Groups*

Since these must be located at termini, standard chemical symbols should be used. These adjoin the appropriate hyphen (for phosphate; *cf.* N-3.1). Examples, in addition to any above [4]:

(CNEt)-; -(CNEt)	5′-cyanoethyl; 3′(or 2′)-cyanoethyl
(MeOPh), (Bzl), (Ph)	anisyl, benzyl, phenyl, with appropriate hyphen.

▲   N-4.4. *Locants and Multipliers*

Multipliers, when necessary, are indicated by the usual *subscripts* [3,8,11]; thus -m$_2$A- signifies a dimethyladenosine residue, neither methyl being at the 2′-*O* position (see N-4.2.1). Locants are indicated by *superscripts*; thus -m$_2^6$A- indicates an N^6-dimethyladenosine residue [ribosyl-6-(dimethylamino) purine], -ac^{4}C indicates an N^4-acetylcytidine, -m$_2^1$,^{6}A- or m^{1}m^{6}A a 1,N^6-dimethyladenosine, etc. [3,8,11]. Utilizing the convention of N-4.2.1, we can write -m$_2^6$Am- for the 2′-*O*-methyl-N^6-dimethyladenosine residue. Other examples are s^{2}U for 2-thiouridine and h$_2^{5,6}$U for 5,6-dihydrouridine (but see the alternates available in N-3.2.1 and N-4.1, namely ^{2}S, and hU or D, respectively; the locants and/or multipliers may be included in the definition). The prefix "di", which has no place in chemical symbolism, should not be used; subscript

numerals suffice. The prefix 2′-*O*-Me [is best replaced by the suffix m (see N-4.2.1), especially when other substituents must be placed before the nucleoside symbol. Thus 2′OMe6Me$_2$A is better symbolized as m$_2^6$Am; similarly, 2MeS6iPeA becomes ms^{2}i^{6}A.

In presenting several homologous sequences, it is often desired to keep the capital letters representing nucleotides one below another. The presence of modifying symbols may interfere with such a presentation. One way of meeting this situation is to place the *prefixes* (including locants and multipliers) directly *over* the capital letter they modify, and to place the *suffix* (usually m for 2′-*O*-methyl) as a right-hand superscript (see also comment ii in N-4.1), *e.g.*, $\overset{m}{\underset{6}{A}}$; C^m.

Examples of this usage exist [3,7,8]. When so placed, smaller letters and/or numbers may be used to advantage [4,8]. Such positioning is consistent with the rules regarding designation of functional groups and their substituents in peptides [5,9].

REFERENCES

1. *Eur. J. Biochem.* 1 (1967) 259, and elsewhere. Section 5 appeared in *Biochim. Biophys. Acta*, 108 (1965) 1.
2. Holley, R. W., Apgar, J., Everett, G. A., Madison, J. T., Marquisee, M., Merrill, S. H., Penswick, J. R., and Zamir, A., *Science*, 147 (1965) 1462.
3. Holley, R. W., *Progr. Nucl. Acid. Res. Mol. Biol.* 8 (1968) 37.
4. Kössel, H., Büchi, H., and Khorana, H. G., *J. Amer. Chem. Soc.* 89 (1967) 2185.
5. *Eur. J. Biochem.* 3 (1967) 129, and elsewhere.
6. Richards, G. M., Tutas, D. J., Wechter, W. J., and Laskowski, M., Sr., *Biochemistry*, 6 (1967) 2908.
7. Woese, C. R., *Progr. Nucl. Acid. Res. Mol. Biol.* 7 (1967) 107.
8. *Handbook of Biochemistry* (edited by H. A. Sober), Chemical Rubber Co., Cleveland, Ohio, second edition 1970.
9. *Eur. J. Biochem.* 1 (1967) 375, and elsewhere. Revision in preparation.
10. Zachau, H. G., Dütting, D., and Feldmann, H., *Hoppe-Seyler's Z. Physiol. Chem.* 347 (1966) 212; *Angew. Chem.* 78 (1966) 392; *Angew. Chem. Int. Ed. Engl.* 5 (1966) 422.
11. Harada, F., Kimura, F., and Nishimura, S., *Biochim. Biophys. Acta*, 195 (1969) 590.
12. Michelson, A. M., Massoulié, J., and Guschlbauer, W., *Progr. Nucl. Acid Res. Mol. Biol.* 6 (1966) 83.
13. Inman, R. B., and Baldwin, R. L., *J. Mol. Biol.* 5 (1962) 172.
14. Ts'o, P. O. P., Rapoport, S. A., and Bollum, F. J., *Biochemistry*, 5 (1966) 4153.
15. Felsenfeld, G., and Miles, H. T., *Annu. Rev. Biochem.* 36 (1967) 407.
16. Ho, N. W. Y., and Gilham, P. T., *Biochemistry*, 6 (1967) 3632.
17. Feldman, M. Ya, *Biochim. Biophys. Acta*, 149 (1967) 20.

All Tentative Rules and Proposals of the IUPAC-IUB Commission on Biochemical Nomenclature (CBN) are available from Waldo E. Cohn, Director, NAS-NRC Office of Biochemical Nomenclature, Oak Ridge National Laboratory, P. O. Box Y, Oak Ridge, Tennessee 37830, U.S.A.:

Abbreviations and Symbols for Chemical Names of Special Interest in Biological Chemistry [see *Eur. J. Biochem.* 1 (1967) 259],

Abbreviated Designation of Amino Acid Derivatives and Peptides [see *Eur. J. Biochem.* 1 (1967) 375],

Rules for Naming Synthetic Modifications of Natural Peptides [see *Eur. J. Biochem.* 1 (1967) 379],

Nomenclature of Vitamins, Coenzymes and Related Compounds: Trivial Names of Miscellaneous Compounds of Importance in Biochemistry, Nomenclature of Quinones with Isoprenoid Side Chains, Nomenclature and Symbols for Folic Acid and Related Compounds, Nomenclature of Corrinoids [see *Eur. J. Biochem.* 2 (1967) 1].

The Nomenclature of Lipids. A Document for Discussion [see *Eur. J. Biochem.* 2 (1967) 127]; amendments [*Eur. J. Biochem.* 12 (1970) 1].

Abbreviated Nomenclature of Synthetic Polypeptides (Polymerized Amino Acides) [see *Eur. J. Biochem.* 3 (1967) 129]; correction [*Eur. J. Biochem.* 12 (1970) 1].

The Nomenclature of Cyclitols [see *Eur. J. Biochem.* 5 (1968) 1].

A One-Letter Notation for Amino Acid Sequences [see *Eur. J. Biochem.* 5 (1968) 151].

The Nomenclature of Steroids [*Eur. J. Biochem.* 10 (1969) 1]; corrections [*Eur. J. Biochem.* 12 (1970) 1].

Abbreviations and Symbols for Nucleic Acids, Polynucleotides and their Constituents [this document].

A document, OBN-5, describing the (American) NAS-NRC Office of Biochemical Nomenclature, and listing other rules affecting biochemical nomenclature is available from its Director, Dr. Waldo E. Cohn [see also *J. Chem. Doc.* 7 (1967) 72].

Note the following corrections in this Appendix:

N-1.1, line 3: *omit* (AtetraP). Symbols, line 11: *for* mononucleotides or nucleotides, *read* mononucleotides and nucleosides. Footnote 9, line 1: *for* Poly AU and AU, etc., *read* Poly AU and poly A+U, etc.

Suggested Bibliography
for Advanced Reading

J. D. Watson (1965). *Molecular Biology of the Gene*. W. A. Benjamin.
The great classic of modern molecular biology.

J. H. Spencer (1972). *The Physics and Chemistry of DNA*.
W. B. Saunders Company.
Large emphasis on the chemistry and sequence of nucleic acids.

V. A. Bloomfield, D. Crothers, and I. Tinoco, Jr. (1973). *Physical
Chemistry of Nucleic Acids*. Harper and Row.
A textbook which includes many advanced techniques and
mathematical treatments.

J. Duchesnes (Editor) (1973). *Physico-chemical Properties of
Nucleic Acids*. Academic Press, 3 volumes.
Multiauthor books containing some of the best reviews on
specific topics.

P. O. P. T'so (Editor) (1974). *Basic Principles in Nucleic Acid
Chemistry*. Academic Press, 3 volumes.
Largely written by Paul T'so, these books are bound to become
the most important detailed reference on nucleic acid chemistry
and structure.

Progress in Nucleic Acid Research and Molecular Biology. 15
volumes published so far (about one a year) under the
editorship of W. C. Cohn. Largely referenced in the present
book, the reviews in *"Progress"* are always the newest accounts
by the best specialists in the field.

Index